T5-BCA-388

Dairy Modernization

Dairy Modernization

Roger W. Palmer

THOMSON
DELMAR LEARNING

Australia Canada Mexico Singapore Spain United Kingdom United States

Dairy Modernization
Roger W. Palmer

Vice President, Career Education Strategic Business Unit:
Dawn Gerrain

Director of Editorial:
Sherry Gomoll

Acquisitions Editor:
David Rosenbaum

Developmental Editor:
Gerald O'Malley

Editorial Assistant:
Christina Gifford

Director of Production:
Wendy A. Troeger

Production Manager:
Carolyn Miller

Production Editor:
Kathryn B. Kucharek

Director of Marketing:
Wendy Mapstone

Marketing Specialist:
Gerard McAvey

Cover Design:
Joe Villanova

Printed in Canada
1 2 3 4 5 XXX 08 07 06 05 04

For more information contact Delmar Learning,
5 Maxwell Drive, PO Box 8007, Clifton Park, NY 12065-2919.

Or you can visit our Internet site at
http://www.delmarlearning.com.

Library of Congress Cataloging-in-Publication Data

Palmer, Roger W. (Roger William), 1943.
Dairy modernization / by Roger Palmer
p. cm.
Includes index.
ISBN 1-4018-4171-6
1. Dairy processing. I. Title.

SF250.5.P25 2005
636.2'142—dc22 2004058447

NOTICE TO THE READER

Publisher does not warrant or guarantee any of the products described herein or perform any independent analysis in connection with any of the product information contained herein. Publisher does not assume, and expressly disclaims, any obligation to obtain and include information other than that provided to it by the manufacturer.

The reader is expressly warned to consider and adopt all safety precautions that might be indicated by the activities herein and to avoid all potential hazards. By following the instructions contained herein, the reader willingly assumes all risks in connection with such instructions.

The Publisher makes no representation or warranties of any kind, including but not limited to, the warranties of fitness for particular purpose or merchantability, nor are any such representations implied with respect to the material set forth herein, and the publisher takes no responsibility with respect to such material. The Publisher shall not be liable for any special, consequential, or exemplary damages resulting, in whole or part, from the readers' use of, or reliance upon, this material.

Contents

Chapter 2 Strategy Development 15

Chapter 3 Facility Planning 27

Chapter 4 Animal Handling Needs 51

Chapter 5 Animal Housing Options 61

Chapter 6 Animal Housing Features 85

Chapter 17 Acquisition of Products and Services 243

Chapter 18 Fitting the Pieces Together 255

Preface

The dairy industry throughout the world has experienced significant changes, resulting in fewer but larger dairies. New technologies allow managers to successfully operate larger dairy herds. "Dairy modernization" typically refers to existing dairy producers changing from one type of production system to another. For example, producers move from a "traditional dairy" system, in which cows are housed and milked in a stall barn, to a system that may include freestall housing, TMR (total mixed rations), and milking parlor. No single system is best for everyone; therefore, producers must understand the available options and evaluate the merits of each for their operation.

The profitability of a business directly influences the quality of life of its owners and workers. Profits can be used to purchase facilities, equipment, and services, which improve working conditions and support family living. Since family living expenses constantly increase, the number of animals or the profit per animal must increase to support growing family needs. Increasing product value or decreasing production costs can influence profit. Modern technologies allow producers to enhance labor efficiency, increase profits, and improve quality of life for both dairy owners and workers. Quality-of-life enhancements help preserve health and safety and often lead to better working conditions, such as more time away from the farm.

These same modern technologies, however, often require larger herds—to decrease the investment per animal and better utilize assets. The optimal herd size varies with the operator's goals and available resources. Each producer must select and incorporate technologies that allow milk production—now and in the future—at a competitive price, and choose the management system and herd size that best provide a profitable and sustainable business.

This text covers those subjects relating to dairy modernization that a producer and decision-support staff must understand when considering building a new dairy or modernizing an existing dairy. Previously, no text book gave comprehensive coverage of the areas involved; this book was written to fill that gap. It is designed for use as a textbook in the classroom and as a reference guide for producers and those in the industry who work with them during the decision-making process. Because it gives comprehensive coverage of the available options and the advantages and disadvantages of these options, it is ideal for bankers, builders, consultants, and others to use as a training tool and reference guide. The dairy industry is constantly evolving—new equipment, facility, and management options are being developed all the time. This book will give readers the basic understanding that will allow them to assimilate new ideas and approaches as they are developed and to form the basis for decision making now and in the future.

Features of This Edition

This text provides comprehensive information on a full range of dairy management issues, including evaluating an operation, strategy development, business and facility planning, animal handling and housing, freestall barn and bedding options, site selection, and milking centers. It also covers the feeding of the dairy herd, manure handling, animal acquisition, heifer raising, labor requirements and scheduling, labor management, record keeping, contracting for services, and, finally, fitting all the pieces together.

Each chapter begins with a list of major objectives and a list of key terms that will be encountered. Each key term is highlighted in bold type when it is first used. Review questions at the end of each chapter provide an opportunity to test comprehension of the chapter objectives.

Supplements

A comprehensive on-line supplement to the text includes additional materials (such as essay questions, etc.). To view the online supplement's Instructor Resources, go to <http://www.agriculture.delmar.com>, click on Instructor Lounge, and follow the steps displayed.

Use of the Text

This text provides the framework for all or part of a dairy herd management class and provides the information needed to teach students the planning process. On-line case studies and field trips can be used to enhance student understanding and to tailor the class to local conditions. The development of exercises using computer-based decision aids and current local benchmark databases will help the student apply the principles presented in the text.

This text is also a valuable tool for the training of industry support people who will work with producers considering the modernization of their operations. Reading the text, answering chapter review questions, and consulting the on-line case studies will provide the trainee with an understanding of industry terminology and facility and management systems options.

About the Author

Roger W. Palmer (B.S. in Mathematics, M.S. in Animal Genetics, Ph.D. in Dairy Management and Agricultural Economics) is an Associate Professor in the Dairy Science Department at the University of Wisconsin. As the department's Dairy Systems Management Specialist, Dr. Palmer teaches dairy herd management courses. His extension activities center around modernization of the dairy industry, and his research efforts are directed toward the evaluation and improvement of dairy facility designs, dairy herd management protocols, and cow comfort issues. Dr. Palmer is the author of numerous articles for dairy publications, including *Hoard's Dairyman, Midwest Dairy Business, Dairy Herd Management, Holstein Science Reports,* and the *Wisconsin Agriculturist,* and several booklets and peer-reviewed scientific articles published in the *Journal of Dairy Science* and the *Professional Animal Scientist.* More than 10,000 copies of his *Dairy Modernization Planning Guide* (2001, Midwest Dairy Business Magazine) have been distributed (in English and Spanish versions) to dairy producers and industry support personnel.

Dr. Palmer travels extensively, both domestically and internationally, consulting on dairy herd facility design and herd management issues. He has been on the Holstein Association's International Consulting Staff since 1981 and has evaluated dairies and given presentations in Russia, Ukraine, Taiwan, Hungary, Egypt, France, Panama, Switzerland, and Czechoslovakia, for the Holstein Association, Purebred Dairy Cattle Association, American Soybean Association, International Executive Service Association, and the USDA Foreign Agricultural Service. His overall interest centers around the business management of dairy farms. He has presented numerous papers at local, regional, national and international meetings, including the European Association of Animals Production Meetings in Zurich (1999) and Rome (2003); the Fifth International Dairy Congress in Panama (1999); and the Western Canadian Dairy Seminar in Red Deer, Alberta, Canada (2002).

Dr. Palmer has many years of industry experience relating to herd expansion, computerized record keeping, and personnel management issues. Before joining the University, Dr. Palmer worked with Purina Mills, Inc.; American Breeders Service; Wisconsin DHIA; and the U.S. Air Force; and managed a Wisconsin dairy farm. He has extensive experience in the development of new dairies and the modernization and expansion of existing dairies of all sizes.

Acknowledgments

This book is dedicated to my father, William S. Palmer, who helped develop my interest in dairy farming, and my wife, Gloria, who has supported me and my efforts throughout the years.

The author would like to express his appreciation to all the producers who have allowed him to photograph their operations as examples to be used in this text, and all his fellow workers who have provided information and support in the development of this work. He would especially like to thank Dennis Armstrong, Dr. Brian Holmes, Dr. David Kummel, Dr. James Smith, and Liz Uhr.

Chapter 1

Evaluating Your Current Operation

KEY TERMS

benchmark databases
debt-per-cow
enterprise
freestall barn
full-time-equivalent (FTE)
milking parlor
modern dairy management system
net farm income
net returns per cow
return-on-assets
return-on-investments
stanchion barn
tie-stall barn
total mixed rations (TMR)
traditional dairy management system

Objectives

After completing the study of this chapter, you should be able to

- identify the key components of an existing dairy's production system, management system, and monitoring system.
- evaluate these components to determine the efficacy of the dairy's site, facilities, equipment, labor force, and management.
- understand the strengths and weaknesses of the dairy's overall operation.
- decide if modernization is a viable option.

What's It All About?

The role of the dairy manager is to plan strategically, and to direct resources in a way that leads to a profitable and sustainable dairy enterprise. Management is the process of decision making, and has three major functions: planning, implementation, and control. The development over the past few decades of new production-enhancing technologies has led to rapid increases in herd size and milk production levels. The challenge to the manager of the modern dairy herd is to economically achieve high milk yield without sacrificing animal health and welfare, quality of the environment, or human safety.

Dairy managers must constantly monitor their operations to ensure that their facilities (production system), procedures (management system), and record keeping (monitoring system) support both the short-term and long-term goals of their businesses. The dairy producer must provide facilities that support cow comfort and labor efficiency at a competitive price. Management procedures must support a profitable production level while maintaining animal health and preserving the environment. Record-keeping systems, which quantify the operation's productivity and profitability, must be accurate, timely, and cost effective.

Dairy managers must also have a long-term operating plan that supports the lifestyle, retirement income, and profitability required by the owners and operators; this plan must also conform to the environmental and animal-welfare concerns of society. The dairy industry is going through a major transition, as dairy operators implement the latest technologies to decrease their investment per animal and increase labor efficiency while meeting the objectives imposed by society. The dairy manager must evaluate differing strategies, then prioritize and implement the changes that will maximize the efficiency of the operation and ensure its long-term viability.

When considering a change to an existing operation, a dairy manager should answer the following questions:

Where are we today?

Where do we want to go?

Which strategy is the best to achieve this?

How can that strategy be implemented?

How can the process be monitored to ensure that it is on track?

Where Are You Today?

In evaluating the status of your current dairy operation, you must first ask, "What is the status of the industry and where is it going?" Then you must ask, "How does my dairy compare to the industry, and how effective will my operation be in the long term?"

FIGURE 1-1 This dairy barn shows that modernization is not new. Dairy producers have historically increased herd size as they have adopted new technology that allows them to manage larger herds using more labor-efficient methods of handling animals.

What Is Happening in the Industry?

A quick look at the dairy industry shows a trend to fewer but larger and more productive herds. In the 40-year period from 1955 to 1995, the number of dairy farms in the United States decreased from 2.7 million to 137,000, while the average herd size increased from 8 to 69 cows per herd. Milk production per cow increased from less than 6,000 pounds per cow per year to over 16,000. What has happened to your herd size, production level and profitability?

Expansion is the term often used to describe this trend. Dairy producers have just participated in the modernization of their industry, made possible by new technologies. Figure 1-1 is a classic example of the type of facility that exists in many parts of the country; here we see a dairy that grew over time from approximately 10 to 80 cows. Each stage in this operation's development was characterized by the implementation of new technology that made farm work easier and labor more efficient. Over the same period of time, crop planting and harvesting progressed from hand labor and horse power to the use of tractors and machinery with continuously increasing capacities. Milking-cow housing changed from **tie-stall** or **stanchion barns,** in which cows are tethered in stalls, to **freestall barns,** in which cows are confined in pens and allowed to walk to feed bunks to eat, water tanks to drink, and stalls to lie down. Feeding went from hand feeding of loose hay

FIGURE 1-2 This two-row drive-through barn design is inexpensive and easy to ventilate.

to mechanical feeding of grain and forage to drive-by feeding of **total mixed rations** (**TMR**). The milking process changed from hand milking to bucket milking machines to pipeline milkers to pit parlors capable of milking a few hundred cows per hour.

This continuing modernization caused the expansion of individual herds. Herds increased because we had the tools to efficiently manage larger herds. Figure 1-2 is an example of how a single-family farm modernized its operation. Before building the new 110-cow freestall barn and double-6 pit parlor, the producer had a 40-cow stanchion barn. This dairy continues to be a single-family operation, but by investing in new equipment and buildings, the family increased the cow numbers and profitability of the operation, as well as their own quality of life.

As you read this book and think about potential changes to your operation, remember that no two dairy operations are exactly alike and that there are several different dairy management systems to choose from. Cows can be housed in open lots or in stanchion, tie-stall, freestall or loose-housing barns. Feed can be harvested by the animal (grazing option), or some portion of her feed can be stored and delivered to her. Cows can be milked where they are housed with bucket milkers or pipeline milkers (known as a **traditional dairy management system**) or taken to a milking facility to be milked (a **modern dairy management system**). Manure and feed can be handled and stored in a variety of ways. Operations may include just one **enterprise** or several: milking cows, raising heifers, cropping, steers, and so on. Operations may be well financed or in financial trouble.

What Does Your Operation Look Like?

To choose the best strategy for your dairy, a sound understanding of your operation's current status is essential. A review of the following areas can provide this information.

Production Performance

DHI (dairy herd improvement) records, computerized herd production records, and milk shipment receipts should be reviewed to determine the quality and quantity of milk produced. How does your production performance compare to local herds with comparable resources? How do your amount of milk shipped and your somatic cell count vary over time? Do you know the exact number of cows milked each day, so that an accurate daily tank average can be calculated and used to monitor the herd?

If you experience substantial changes in bulk tank shipments, do you review the herd's calving patterns and culling rate to understand why? Do you have a high culling rate? If so, is it an indicator of poor herd health, or of aggressive culling to achieve a high milk output per milking animal?

Do you know the average days-in-milk of the herd and of individual groups of animals, and do you use this information to identify herd man agement problems relating to nutrition, breeding, and so on? Do you monitor herd somatic cell counts to give insight into the herd's overall udder health?

Current Financial Performance

Do you have a method of keeping financial records that accurately depicts the profitability, liquidity, and solvency of your operation? Do you monitor indicators of profitability, such as **net farm income, return-on-assets,** and **return-on-investments?** Are you happy with the level of returns received for your investments?

Liquidity measures based on your operation's current expected income and expense indicate your operation's short-term financial situation. Do you have a current income statement? Does it show that you have significant accounts payable, or are you currently experiencing cash flow problems? Solvency measures, such as **debt-per-cow,** total net worth, and percent equity, can be found on your balance sheet and indicate the long-term financial strength of your operation. Does your balance sheet support the type of investment you are considering? Remember that lenders normally want a dairy operation to have a maximum debt-per-cow of $3,000–3,500 and a minimum equity position of 35–50 percent after any major change.

Growth of Assets and Equity over Time

A quick review of your operation's balance sheets for the past few years will indicate the growth and financial strength of the business over time. A consistent increase in total assets is a good indicator of your business's growth, and a consistent increase in the operation's net worth is an indicator of your operation's profitability over time. Remember that these growth trends often are viewed positively by lenders, investors, and suppliers, and will have a major impact on your ability to attract capital.

History of Animal Health Problems

A review of your dairy's veterinary bills and health records can give insight into the animal comfort and animal management status of the operation. **Benchmark databases** are normally available from university and industry sources to help you define normal expected veterinary costs and health problem frequencies. Do you know the average percentage of your herd that has had ketosis, displaced abomasums, and so on? If severe health-related problems have been experienced, your current vaccination and treatment protocols should be reviewed.

Quantity and Quality of Records

As operations increase in size or production intensity, the importance of adequate records increases. Low-cost personal computers and software programs make record keeping easier than in the past. Both financial and production record-keeping systems are needed. Manual records can provide necessary information to you and your support staff, but computerized records are much easier to analyze. Is your record-keeping system adequate to support the decision-making needs of your current and proposed operation? Does someone in your operation have the requisite record-keeping skills and interest?

Efficiency of Operation

The efficiency of an operation is often described by measures of labor efficiency and profitability. Milk per worker is often used, where worker numbers are calculated using a **full-time-equivalent** (**FTE**) of 50–55 hours per week. Current dairy operators are encouraged to achieve one million pounds of milk per FTE per year, and some of the more efficient modern dairies surpass this measure considerably. Do you know these values for your operation, and are you happy with them?

Cost per hundred pounds of milk sold is another common measure and reflects an operation's efficiency of production. **Net returns per cow** standardize the profitability of the operation on a per-cow basis. Both of these values should be looked at in combination with total net returns to ensure that there is sufficient income to cover family living expenses and so forth.

Efficiency of Individual Components

A dairy operation can be evaluated by the activities performed. What are the relative efficiencies of the milking, feeding, animal handling, and cropping activities of your operation? The biggest labor component of a dairy is milking. How many people are involved in the milking process, and how long does it normally take? Use these values to calculate the number of cows per worker hour. Tie-stall barns have been shown to average 30–35 cows per milker hour, whereas large automated parlors can achieve one hundred or more. How does your dairy compare?

Analysis of Enterprises

Dairy farms differ significantly in their level of specialization and may have only a single enterprise (milk cows) or several (milk cows, heifers, crops, steers, etc.). To choose the best strategy for your operation, you must know the relative profitability of each enterprise. Although separation of costs by enterprise is often difficult, it is worth trying to estimate these values to determine how best to use existing assets. Do your plans have the potential to maximize the use of available funds?

Feed Quality, Inventory, and Handling Practices

High-quality forage is a key ingredient for a high-producing, profitable dairy. Take time to review the quality of your dairy's feedstuffs, feed storage facilities, and handling practices. Ration-balancing and feedstuff analysis reports should be reviewed by an expert to help you ensure that your feedstuffs are of sufficient quality and that rations for different animal management groups meet their nutrient needs.

What Do Your Existing Facilities Look Like?

Since facilities are a key to the effectiveness of an operation, significant time and effort should be spent evaluating your current facility to determine its future value and utility.

Current and Future Capacity

What are the current and potential capacities of your facility? Evaluate each housing unit, milking facility, and other potentially useful building (machine sheds, hog barns, etc.). Existing buildings often can be converted to freestall housing by opening sidewalls and adding or changing feeding and resting areas. The ability to overstock facilities should be evaluated. Ventilation, feed space, milking capacity, and animal handling capabilities can influence the number of animals a building can hold.

Physical Condition: Ability to Repair, Expand, Modify

When evaluating existing facilities, pay attention to their structural aspect (Figures 1-3 and 1-4). What are the conditions of concrete floors, frame structures, and roofs? The best use for a structure depends on its size, location, and condition. If barns of pole construction are to be used, the

FIGURE 1-3 Modernization of existing facilities can be very expensive, and cost overruns are frequently encountered. It pays to hire an experienced construction engineer to determine what is required to repair and update existing facilities.

FIGURE 1-4 Use of an existing building should be considered if it is structurally sound and located on a site that allows for the long-term growth needs of the operation. Here an existing dairy barn and milk house were used, but an inexpensive holding area was added to provide proper ventilation and aid in cow comfort.

location and spacing of poles is important in deciding the best use of the structure. Before retrofitting an existing facility, consider retrofitting costs, useful life of the building, and the building's usefulness compared to a new structure. Experience has shown that retrofitting existing buildings is often more costly and less satisfactory than originally estimated.

Growth Potential

If an existing operation is to be expanded, its site is an important consideration. If the site does not support the long-term growth you anticipate, then a new site should seriously be considered. You might be tempted to build a new facility at the old farmstead, knowing that there is no future growth potential but assuming it will be used later for dry cows or heifers. Although this strategy may make sense in some instances, too often it results in inefficient use of resources and delays the move to a better site with its associated profits.

Ability to Group Animals

Do your existing facilities provide for efficient housing, management, and milking of cows according to their production, health, and reproductive status? If not, could it be modified to do so? As production levels of herds increase, it becomes more important to house and handle separately the far-off dry, pre-calving dry, maternity, post-calving, sick, and milking cows.

Labor Efficiency

What are the labor requirements of your existing facilities and any proposed modifications? Initial cost should be combined with ongoing labor costs to arrive at an annualized cost per cow for any changes proposed.

Cow Comfort

Cow comfort is a key element leading to high production of a dairy herd. Advances in genetics, nutrition, and production technology have led to higher herd production levels; currently, several herds in the United States have rolling herd averages for milk in excess of 30,000 pounds per year. The impact of the facilities chosen on cow comfort is critical to the expansion decision.

Worker Comfort and Safety

A primary objective of most dairy farms is to provide comfortable and safe working conditions for the owner(s), resident family members, and employees. Large labor turnover rates and the departure of family members are often the results of an undesirable working environment.

Feed Storage and Handling

Large and small herds require different types of feed storage. Upright silos make sense for small operations, whereas flat feed storage makes sense as

FIGURE 1-5 Manure is often collected in freestall barns by skid steers that push the manure to a central location. Manure drops through a floor opening into a flume that leads to permanent manure storage outside the freestall barn.

herd sizes increase. As herd sizes grow, it becomes a real challenge to determine when and how to use existing feed storage systems. Total mixed rations have proven to be beneficial for herds of all sizes but often are not used on small farms because of costs—for equipment, for new storage facilities, and for transport of ingredients to a mixer and the TMR to the animal. For larger herds portable TMR mixers are used to mix and deliver feed to individual groups of animals. Any new or modified facility should be designed to accommodate TMRs and drive-by feed delivery.

Manure Management

What type of manure handling system currently exists (Figure 1-5)? What is its capacity? Where is it located? Is it approved under current zoning regulations? Will this system be usable for all or part of the operation's manure storage needs, and if so, for how long?

What Does Your Existing Equipment Look Like?

To evaluate existing equipment, take into account its age, condition, and usefulness. Expanded dairies sometimes attempt to get by with existing equipment. This will often lead to poor-quality feeds; either the equipment breaks down with the excessive use, or insufficient harvesting capacity results in delayed harvesting.

What Are Your Land Resources?

The ultimate size of a dairy operation is limited by its ability to provide forage and to dispose of manure. When making a major investment in a dairy, you must consider the amount of land you own or rent, and how much you can expect to find in the future. The current nutrient loading of the land should be determined so that animal numbers are balanced with the ability to dispose of the manure. Consider also the productivity of land, because highly productive land is normally more profitable in the long run. Another aspect of long-term dairy planning is the relative cost of land ownership versus rental costs. The decision to own or rent land is complicated by the risks associated with renting, the potential inflation possibilities with ownership, capital gains implications, the expected current returns from an investment in land versus in other dairy-related assets, and how the decision complements your long-term growth path.

What Is the Quality of Your Existing Herd?

When you evaluate your dairy herd, consider its genetic level and animal condition. Different breeds of cattle have different abilities for milk and milk component production. The level of artificial insemination (AI), participation in breed development programs, and your sire selection philosophy indicate the genetic potential of your herd. If you are on DHI testing and report AI sire usage, your processing center probably summarizes the genetic trends within your herd. A visual inspection of your animals for cleanliness, body condition, size, and health problems (feet, legs, udders, etc.) will indicate the cow comfort of your facilities and your team's animal-management skills.

What Is the Quality of Your Existing Labor Force?

How many family members and hired workers currently are (or want to be) involved in the operation? Define the age, interests, skills, and compensation needs of each member of this existing labor pool. One of the objects of the planning process is to create an operation large enough to fully utilize the skills and interests of these workers as well as satisfy their income needs. Next, analyze the local area to determine the potential for additional employees and the compensation level offered by other industries in competition for that labor supply.

What Is the Quality of Your Management Team?

A management team analysis is one of the most critical aspects of the evaluation process because, first and foremost, a dairy operation is a business and must be run by competent management. Several observations should reveal the capabilities of your existing management team. What is the physical appearance of the operation—its neatness and level of maintenance—and is there any evidence of obvious waste?

What is the ability of your existing team to organize, set priorities, and direct activities? Consider past actions and justifications to determine if the necessary decision-making ability, level of industry awareness, and motivation exist in your management team. A successful dairy manager must have the skills to work with people. How much experience with hired help have you had, and how successful have you been?

What is the personality type of each management team member? Are they introverts, who like to work alone, or extroverts, who like to work with people? Do the managers have an interest in interacting with employees and sales and support people? Since record keeping is such an important part of the management process, it is extremely important to determine the managers' experience and interest in record keeping and analysis. Successful management of a modern dairy requires knowing why records are kept and how to use them for decision making. Does your management team understand your production record-keeping system, as shown by each team members' knowledge of key herd-health, production, and reproduction values? Their understanding of the key economic measures that must be maintained and monitored indicates their competency with financial records.

What Are the Major Weaknesses of Your Operation?

All of the preceding observations should be consolidated so that you can identify the weakest elements in your operation. This list should be an important part of any future planning. Will the changes that you are contemplating fix existing problems or make them worse?

SUMMARY

Dairy producers must monitor their operation constantly to ensure that they are positioned to be completive in the long term. For many existing producers, modernization and expansion are options worth exploring. Many new technologies can aid producers in improving the efficiency and profitability of their operations. Each producer's circumstances and needs are unique, and the process of determining a long-term strategy that will satisfy the operator's goals should not be taken lightly. This chapter has set forth the aspects of current operations that producers should evaluate in pursuing this process.

CHAPTER REVIEW

1. What dairy activity requires the most labor?
2. List the four enterprises commonly associated with a dairy herd.
3. What are two key indicators of cow comfort?
4. What is a balance sheet used to evaluate?
5. High-producing, well-managed herds in the United States currently have rolling herd averages at or near what production level?
6. What does a "production system" consist of? A "management system"? A "monitoring system"?
7. Write a paragraph explaining dairy modernization. What is the ultimate goal? What operational systems are affected?

REFERENCE

Farm Financial Standards Board. (1995). *Financial guidelines for agriculture producers.* Norwalk, CT: Author.

Chapter 2

Strategy Development

KEY TERMS

biological subsystem
business structure
buy-sell agreements
C-corporation
cash-flow projections
economic subsystem
estate plan
financial liability
general partnership
hybrid organization
implementation plan
limited liability corporation (LLC)
limited partnership
organic products
S-corporation
social subsystem
succession planning
technical subsystem

Objectives

After completing the study of this chapter, you should be able to

- identify the biological, economic, social, and technical subsystems within an existing dairy operation, and their interrelationships.
- develop a comprehensive modernization plan that supports the operation's short- and long-term goals.
- formulate a list of current and future objectives for the operation that all stakeholders can agree upon.
- incorporate these objectives into a realistic business plan with the assistance of legal, financial, and business development professionals.

After evaluating the overall status of your operation and its components, it is time to look ahead and answer the question, "Where do we go from here?" You should ask many questions, look at many other operations, listen intently, takes notes and photographs, and above all keep an open mind about what you observe. Research your industry to determine current trends, to understand how the industry is changing, and to decide if and how you should change your operation.

Will You Need to Increase Herd Size If You Modernize?

As mentioned, the dairy industry is going through a major transition, with herd owners investing in new technologies that allow them to operate with a lower investment per cow, to improve their labor efficiency, and to improve their quality of life. This modernization of facilities often leads to herd expansion because

1. labor efficiency gains allow the same size of labor force to handle more cows
2. it is necessary to spread improvement costs across more animal units
3. economies of scale are gained
4. advantages are gained by employee specialization

How Do You Make a Business?

One of the first decisions you must make is which type of dairy system to select. In the United States, most operators choose the large confinement-based model with freestall or drylot housing, whereas producers in countries with warmer climates choose the pasture-based model. A third system, which is gaining popularity in Europe and North America, is robotic milking. Robotic milking systems have proven to be reliable and to cut the milking labor requirement. Each system has advantages, and you should evaluate the feasibility of each system based on your circumstances.

The role of the dairy business manager is complex because many different subsystems and their interactions must be considered. The **biological subsystem** involves the cow and proper management of her needs, allowing her to produce large amounts of milk efficiently. The **economic subsystem** requires the manager to exercise cost control when purchasing inputs, and good judgment when marketing outputs. The **social subsystem** involves influencing hired labor to work in a harmonious manner to consistently and efficiently produce a quality product. (Producers making a change from a traditional dairy management system to a larger, more modern system must realize that success will depend less on their ability to manage cows than on their ability to manage employees.) The **technical subsystem** requires selection and use of machines and tools that complement the needs of the overall dairy production system.

The efficiency of a dairy operation is greatly influenced by the design of the facilities used to house and manage the herd. Strict attention to the flows of cows, people, equipment, manure, feed, and air in the design of a facility will directly contribute to the profitability of the operation—and its resulting cash flow. The design features selected will influence the dairy managers' options for managing the dairy herd. The manager must simultaneously consider animal and facility management, feed management, personnel management, and financial management issues. These issues must be viewed by the manager in the context of the whole dairy operation, because of the interactions between each area.

The business manager must evaluate each potential management decision with the following set of questions:

How much will it cost?

How much will it return?

What is the likelihood of success?

What is the risk of failure?

How much time is expected between investment and return?

How will performance be measured?

Determine Your Goals

Goals are merely targets or desired outcomes that motivate a manager. Early in the planning process, it is important that all parties with ownership interests in the dairy operation meet to determine the mission and goals of the operation. Participants should define their goals independently before the meeting. They should consider their personal goals and how they might relate to the business's goals. The business's goals should then be defined to incorporate the mutually-agreed-upon goals of the individuals. These goals should form the basis of a long-term plan and be documented in a formal business plan.

Review Your Options

To determine what options exist for you and your dairy, you need to think about the dairy industry and what has happened over the past 50 years. Review the technologies that were introduced and how they affected the industry's average herd size, milk production, labor efficiency, and so on. What has been the effect of milking machines, barn cleaners, bulk tanks, milking parlors, and freestall barns? Review other industries and try to identify how they changed to meet consumer demand. (As an example, at one time gasoline and groceries were sold by small, independent, locally owned gas stations and grocery stores. These have been replaced by convenience stores selling gas and some high-demand grocery items, and warehouse-size grocery stores selling the bulk of one's grocery needs.) In the past, dairy farms included several enterprises (milk cows, heifers, crops, etc.) and were family owned and operated with financing from family equity and conventional debt. Generic products (milk and meat) were sold on impersonal markets. These trends are changing. Some producers are specializing by business enterprise: milking cows, custom cropping, custom heifer raising, and so on. Some are becoming contract growers, marketing tailored products through contracts and strategic alliances with processors. Ownership may include several different investors. These trends are expected to continue because of the large investment of capital and management resources required to implement the large-scale ventures that our modern technologies support.

After conducting these mental exercises, you should create a list of possible strategies. Develop as comprehensive a list as possible, sharing it with support personnel (friends, lenders, consultants, etc.) to get additional ideas that can be used to expand and evaluate the list.

Establish the Critical Factors

A dairy producer must consider many interrelated factors when developing the long-term plan for a dairy. Selection of the correct herd management system, herd size, and facility type will be based on lifestyle choices, the type of product to be marketed, land and labor availability, and so on.

Marketing a generic product will continue to be a viable option, but margins for commodity products cannot be expected to be as large as those for differentiated products. Specializing in high-cheese-yield milk or **organic products** currently can increase product value. Soon dairies will be able to contract with pharmaceutical companies to deliver milk from genetically altered animals, milk genetically designed to yield pharmaceutical products.

The geographic location of the dairy will be less important with time because technologies such as ultrafiltration can decrease shipping cost. This allows dairies to be located near the source of feed and manure disposal, and away from urban centers.

Evaluating the enterprises with which your operation should be involved is a very important part of the process. This evaluation should not limit your options to what has been done in the past or what your neighbors are doing. Consider adding enterprises that complement your operation and dropping enterprises that do not. Some dairy operators successfully manage dairy-related enterprises (such as milk hauling, bedding sales, feed marketing, and hoof trimming) that have proven to increase the overall efficiency and profitability of their operations. (The importance and profitability of raising breeding stock will change as sexing of semen, cloning, and other reproductive techniques become available.)

Consider Both Long- and Short-Term Goals

When contemplating any major change to an existing dairy, it is critical to define the long-term goals of the operation and then to define the steps or phases needed to achieve these goals. An important part of the strategy development process is identifying not only feasible strategies but also different alternatives for implementing them. For example, your short-term goal may be to build a double-8 (D-8) milking parlor and increase herd size to 200 cows, but your long-term goal is to expand this D-8 parlor to a D-12 and milk 700 cows. If, during the planning process, this is not considered, then new buildings and manure storage, may be placed incorrectly for the potential long-term growth of the operation. Since a fully utilized double-8 milking parlor can support a dairy of 500–800 cows (depending on the choice between 2X or 3X milking), consideration should be given to how the site could be used if herd size were increased. In this case the 200-cow herd size would be considered phase I of a long-term strategy.

Evaluate Different Strategies

Once a list of possible strategies has been completed, it is time to select those that are feasible, that meet your objectives, and that coincide with your personal preference. Start by pruning the list of those options that obviously do not meet these criteria.

Next work with the other owners to determine the resources that could be made available (feed, manure rights, labor, investment capital, etc.). At this time it is very important to determine your operation's financial borrowing capacity by working with your lender or a financial consultant. Your financial borrowing capacity and rough cost estimates can help you eliminate more options. At this point, ideally, only a few options remain, and it is time to do a financial feasibility analysis for each.

Realistic **cash-flow projections** based on realistic input costs, production levels, and product values should be developed for a five- or seven-year projection time period. Sensitivity analysis for key variables (such as production level and milk price) should be done to see the effects of changes on these key variables. Monthly cash-flow analysis will help you and your lender identify periods when additional funds may be needed.

The final selection of strategy should then be made by all parties involved. The best decision normally can be found by again asking the following set of questions for each strategy being evaluated.

How much will it cost?

How much will it return?

What is the likelihood of success?

What is the risk of failure?

How much time is expected between investment and return?

How will performance be measured?

Remember, this is a critical time in the planning process, and it is extremely important that participants "buy into" the decision.

Business Planning

Now that you have evaluated the status of your overall operation, defined your goals, evaluated different strategies, and selected a strategy to pursue, you are ready to answer the question "What legal framework is best for my operation?" This is the time to define how you plan to reward owners and family members fairly for their inputs. A sound business, like a good house, needs to be built on a strong foundation. Before any major change to a business begins, the owners should review previous legal and moral obligations. It is especially important with family-owned operations that owners review each family member's contributions to the business and determine if they have been fairly compensated for their inputs. Current and desired ownership and compensation should be defined. This is the time to fix any past inequities and to define an equitable system for the future operation of the business. The size and scope of the changes desired should be defined and proper systems put in place to effect desired changes.

The biggest threat to business is the delicate nature of human relationships. It is fundamental that all participants maintain trust to avoid conflict, which ultimately increases the stress levels of everyone involved. Good communication skills are essential for the successful operation of multiowner operations. This does not happen by itself; everyone involved must work at developing good verbal and listening skills. During this part of the planning phase, it is important to establish the framework that will foster the long-term working relationships desired.

This is the time to solicit the support of attorneys, accountants, and business development professionals to help you understand available legal, financial, and business structure options, and to help you select those that are the best for your individual situation. All of these professionals can help you develop the proper legal framework once you and other owners and family members have agreed on the goals of the operation.

Business Plan Development

The key point of the preceding is that the existing and potential participants in the new or changing business must know what they want to accomplish before attorneys, accountants, or business development professionals can help. Having a well-developed list of objectives will allow these professionals to help you define and evaluate your options.

To facilitate the planning process, a formal business plan should be created. It should clearly define what is being planned and how it will be implemented. A well-written business plan accurately details the owner's understanding of what is planned. It is an effective way to increase the understanding of owners, family members, workers, and lenders involved with the project. If done correctly, it should create and maintain trust among owners—and throughout the organization.

Business Planning Goals and Objectives

The following is a summary of the basic planning goals and objectives of most dairy farm owners.

1. **Financial security**—the confidence that sufficient funds will be available to cover living expenses, health care, and protection of assets for the lifetime of the owner(s) and their spouse(s)
2. **Profitability**—the ability for the business to generate sufficient profits to ensure its financial stability, to generate an acceptable return on investment, to provide income to support an acceptable lifestyle, and to create sufficient wealth to support long-term retirement needs
3. **Quality of life**—working conditions and profitability that allow the owners and workers to have a good balance between time commitments to the dairy and other aspects of their lives
4. **Transfer to the next generation**—a business that can be passed to future generations and continue to be viable after the transfer
5. **Treat children fairly**—a method of distributing assets to on- and off-farm heirs in a way that is equitable and does not destabilize the operation
6. **Avoid income and estate taxes**—minimization of the short-term and long-term tax burden
7. **Limit probate**—minimization of any delays and expenses associated with probate

Business Plan Contents

Your business plan should include several different kinds of information.

1. **Description of current operation:** What assets are currently owned or controlled: cattle, machinery, buildings, and land? Explain the operation's current business structure. Document any real estate, machinery, or livestock owned jointly, and any leases, contracts, or other legal arrangements with other people or organizations. Describe the amount and quality of available labor and management and how it is organized. Explain how the business has been financed and all current debt obligations. Explain how the size and productivity of the operation has changed over time. Include financial information that explains the profitability, liquidity, and solvency of the operation over time. Include any additional information that will help the reader understand the current status of the operation.

2. **Implementation plan:** In your **implementation plan,** define the strategy that has been selected, explain why it was selected, and list the goals and objectives that are to be pursued. Define the proposed business structure and organization. Outline the required facilities and other resources needed, construction and operating budgets, and the financial requirements of the project. (An important element that is often omitted is the contingency plan that will be deployed if some catastrophic event happens.)
3. **Business structure plan:** Include a definition of the **business structure** that has been selected, and copies of all legal agreements associated with the business. The following are some of the criteria that are often used when selecting a business structure.
    - **Control**—owners must decide on the amount of control they want over current operations, and if they wish to retain some level of control after they retire.
    - **Ownership transfer**—often owners want to restrict transfer of ownership interests, to assure that the business will remain in the control of existing owners or planned successors.
    - **Liability**—most owners would like to limit their **financial liability.** Limited liability is an attribute of some organizational structures. It limits the amount of the business's liability that is assumed by an owner to the amount invested or to an amount the owner has personally guaranteed.
    - **Tax implications**—owners need to be concerned with how profits from the business operation or its liquidation are taxed. Different business structures allow cash generated to be retained by the entity or to flow through to the owners, resulting in different income tax liabilities. Avoiding or minimizing taxes when a business is liquidated is also important.
    - **Simplicity**—simplicity aids understanding and often decreases the cost of forming and operating a specific business entity.
    - **Valuation discounts**—sometimes it is desirable to choose a business structure that allows valuation discounts for gifts of business interests. These are often used to transfer assets to junior family members at a reduced cost.

Business Structure Alternatives

The following is a brief review of the different business structures that are available to a dairy producer. It is extremely important to work with a professional when selecting a business structure because the tax and non-tax characteristics need to be fully understood. Remember to consider both the short- and long-term implications and, if possible, calculate the total economic benefit expected from making a given choice. The sole proprietorship business organization is often used if the operation has one owner who has full control of and responsibility for all management decisions. The sole proprietor has the responsibility for all management decisions, is liable for all debt, has no restrictions on the sale of the business, and requires no formal agreements.

Partnerships of two or more owners who share profits and losses can be either general or limited. With **general partnerships,** each partner is liable for all debts, can participate in management, and can bind other partners to contractual obligations. **Limited partnerships** must have at least one general and at least one limited partner. In this case the limited partners have limited liability, as mentioned, and cannot participate in management of the business. General partnership agreements need not be in writing, but limited partnerships must file a written partnership agreement with the secretary of state.

A newer form of business structure is the **limited liability corporation (LLC).** Participants are called members and, as the name of the business structure indicates, members are not liable for the debts of the LLC unless they have personally guaranteed those debts (i.e., loss is limited to the amount paid for the member's interest in the LLC). LLCs are formed by filing articles of organization with the secretary of state, and must have two or more members. LLCs may be taxed as a partnership (often referred to as an LLP) or can elect to be taxed as a corporation.

Corporations are another type of business arrangement, and articles of incorporation must be filed with the secretary of state when formed. Three variations exist, the **C-corporation,** the **S-corporation,** and the cooperative. Owners are referred to as shareholders in corporations or members in cooperatives. Each of these forms requires the owners to elect a board of directors who hire officers to manage the business. In most farm operations, owners are elected to the board and are hired to manage the operation. Shareholders have limited liability. In corporations, stockholders vote according to their ownership percentage, but in cooperatives each member has only one vote no matter the amount owned. S-corporations differ from C-corporations in that income may be taxed at the owners' level, thus avoiding a major disadvantage of the C-corporation, in which income can be taxed twice—once as business profits and again as personal income once profits are distributed to owners.

Hybrid organizations are often used for large operations. Such businesses combine two or more organization types to take advantage of their different characteristics. For example, a C-corporation may be formed to own the operating assets of a large dairy (cows, buildings, and equipment), and an LLC formed to own the land needed for the dairy. In this arrangement, the appreciable assets are held in the LLC, which has favorable rules for handling capital gains, and the C-corporation allows owner-employees' fringe benefits to be treated as operating expenses, which can be deducted by the corporation.

Many references describe in more detail the tax and non-tax characteristics of each of these business arrangements. These can be reviewed if an in-depth understanding is desired, but most often the professionals you hire will help you select the correct organizational structure based on your specific circumstances and desires. In general, LLCs offer the most advantages: lower administrative costs, limited liability, flexibility in distributions, single taxation of income, capital gains benefits, and favorable adjustments on the basis of assets. Partnerships tend to be avoided because they do not offer limited liability. C-corporations and cooperatives have the advantage of tax deductions for fringe benefits paid to employee-owners and are desirable when implementing a multientity or hybrid organization. The cooperative structure is not commonly used for farm businesses because cooperatives normally are limited to a maximum of 8 percent return-on-investments, and any remaining business profits must be distributed based on some form of patronage or "use." When selecting a corporation-type business structure, you must decide if the ability to deduct fringe benefits is worth the extra complexity and administrative costs associated with the corporation structure (or multientity organization). The business structure you select should fairly reward all interested parties for their contribution of capital and labor.

Succession, Retirement, and Estate Plans

Each owner should consider the effect of any agreement on the transfer of the business to another person, on retirement, and on estate plans. **Succession planning** deals with the transfer of the owner's interest in the business at the time of death, retirement, or withdrawal from the operation. Transferring ownership to future generations should begin before the age when death is likely. A business transfer agreement, or **buy-sell agreement,** should be developed to specify exactly how this would be done. Retirement plans should be developed that will provide sufficient income to support the standard of living expected by the owner and spouse. The amount of income needed to support an exiting owner can vary, based on what initiates the withdrawal, the owner's level of nonfarm investments, social security benefits, and so on. Legal agreements that protect the owner's equity and the remaining owner's right to continue the operation are extremely important. A well-developed **estate plan** should define what the descendant wants to happen to owned property. It should balance the needs of the person, spouse, and both on- and off-farm dependent children. It may include bequests, price discounts, or purchase options that make it possible for interested family member to continue the business. Experience has shown that nonfarming family members can often cause undue tension and stress on everyone involved at the time of a death if procedures are not in place to support the owner's desires.

Buy-Sell Agreements

Buy-sell agreements are often included in a business plan to restrict the transfer of ownership interests and determine how owners may leave the business. They can be used to prevent the transfer of ownership to an undesirable third party, and they can be written to discourage early withdrawal of ownership interests, which may negatively affect the business. Often they identify who will purchase an owner's interest in case of death, disability, retirement, or withdrawal. They should be written to treat the remaining owners in an equitable manner and may be used to help assure the continuation of the business. They should specify how remaining owners can buy out inactive members, and establish a method for determining the value of an owner's interest. They can establish provisions for funding the purchase of an owner's interest by other owners; these provisions could include contract conditions, interest rates and down payment levels for notes, or specifications that life insurance will be purchased and used to fund such a buyout. This is an extremely important part of the overall planning process. It is important to define the potential situations that could arise, their impact on the business, and their impact on each owner's options; these should be clearly delineated in this type of agreement.

SUMMARY

The business planning process associated with creating a new dairy, or improving an existing one, is very important. After the owners have agreed on the goals of the business, a business plan should be developed that delineates in detail exactly what is desired and how it will be accomplished. This document should deal with all the important aspects of the business and the owners' relationships. It should be considered a living document in that it needs to be reviewed and updated periodically. It should not only define how the business will be organized and operated but also include procedures and agreements relating to dissolution of the business. Choosing a business structure is a key part of this process, and one that is complicated by the fact that each legal business structure has different characteristics. To understand how each of these will affect your business in the short and long term, you should solicit the help of attorneys, accountants, and business development professionals in selecting the best option for your individual situation.

CHAPTER REVIEW

1. What four factors often lead to herd expansion after modernization?
2. List the four subsystems of a dairy management system.
3. What individuals should be involved in formulating a dairy operation's mission? Identify the steps in developing an operation's goals.
4. What are the four primary factors that must be considered when developing a long-term plan for a dairy?
5. List two key resources that must be evaluated to determine the feasibility of an expansion.
6. List three purposes of a business plan.
7. Imagine that you are assisting a family-owned operation in writing a business plan. Make a list of the parties that should be involved in the business planning process. Make a list of the decisions that need to be made prior to formalizing the plan with a lawyer.

REFERENCES

Harris, P. E. (1997). *Business ownership and management assessment.* Madison: University of Wisconsin, Department of Agricultural and Applied Economics.

Lawless, G., Cropp, R., and Harris, P. E. (1997). *Cooperative ownership compared to other business arrangements for closely-held joint ventures.* Madison: University of Wisconsin, Center for Cooperatives.

Twohig, G. W. (2000, March 7). *Planning for the right business structure.* Handout, PDPW Annual Business Conference, Madison, WI.

Chapter 3

Facility Planning

KEY TERMS

AMS (automated milking system)
corral
DHIA (Dairy Herd Improvement Association)
drylot system
H-style
labor system
large modern confinement system
manure handling system
MIRG (management intensive rotational grazing)
modified H-style
NFIFO (net farm income from operations)
T-style
volume price premium

OBJECTIVES

After completing the study of this chapter, you should be able to

- understand the interdependence between the housing, feeding, manure handling, animal handling, and milking subsystems within an existing dairy operation.
- understand how grazing, confinement, drylot, and robotic dairy management systems utilize these subsystems differently.
- determine which dairy management system works best, given your operation's financial position, goals, geographic location, and existing facilities.
- determine the scope of facility modernization possible at present and in the future.

Now that you have evaluated the status of your current operation, defined your goals, selected a strategy to pursue, and decided on the legal framework under which you will operate, it is time to consider facility-related options. One of the major goals of modernization is to improve the efficiency of the overall operation. The efficiency of an operation is greatly influenced by the design of the facilities used to house and manage the dairy herd. As previously stated, strict attention to the flows of animals, people, equipment, manure, feed, and air in the design of a facility directly contributes to the profitability of the operation and its resulting cash flow.

You must make many decisions when building a new dairy facility or undertaking a major expansion of an existing operation. The design features selected influence the dairy managers' options for managing the dairy herd. Choices relating to milking, housing, manure handling, feeding, feed storage, and animal handling must be made from the many different options that exist for each. Each of the system components selected must be compatible with the others. Before individual components can be selected, you should research and answer the following basic questions:

How much money is available?

Which dairy management system will be selected?

What size and type of milking facility will be used?

How will manure be handled?

How will cows be fed?

What freestall bedding material will be used?

How and where will animals be handled?

Which parts of the existing facilities will be used, and for what purposes?

Knowing the answer to each of these questions provides the information needed to determine the best site for the dairy, the proper arrangement for each of the components, and whether a phased approach is desirable to meet your long-term goals.

How Much Money Is Available?

Before making any dairy modernization buying decisions, you should know your budget constraints. If your modernization is to be financed using current assets for collateral, you can work with your lender or use the following approach to estimate your potential borrowing capacity.

Table 3-1 shows how to determine the approximate loan amount you can expect, based on what you own, your equity position, and your lender's lending guidelines. First, determine the value of all assets currently owned (A), current liabilities (B), and net worth (C) (these can be obtained from a current balance sheet). Since lenders normally require a 35–50 percent

TABLE 3-1 Maximum loan amount determinations ($ US)

Current Status		
(A) Assets	**(B) Liabilities**	**(C) Equity or Net Worth**
$883,000	$100,000	$783,000
Equity % after loan	(D1) Min. % ÷ 100 .35	(D2) Max. % ÷ 100 .50
Maximum assets	(E1 = C ÷ D1) $2,237,143	(E2 = C ÷ D2) $1,566,000
Maximum loan	(F1 = E1 – A) $1,404,143	(F2 = E2 – A) $683,000

equity position after expansion, use .35 and .50 to determine the maximum range of values of assets that can be owned after expansion (as shown in the table). Since you will not sell the assets currently owned, the value of these assets (A) must be subtracted from the maximum-assets value to give an indication of current borrowing capacity. The example in the table shows how a producer with a net worth of $783,000 has a potential borrowing capacity of $683,000 to $1,404,143.

By knowing your potential borrowing capacity and the amount of any money from other sources, you can determine the size and scope of your modernization effort. If funds are not available to obtain all the assets desired or if you are not willing to borrow to your maximum capacity, then a phased approach should be considered. With a phased approach, a complete layout for a larger facility is made, but only a portion is constructed. This phased approach ensures adequate space for future facilities and is designed so that permanent structures are not placed in a way that hinders eventual facility layout.

What Dairy Management System Will Be Selected?

Merriam-Webster's dictionary defines a system as "a regularly interacting or interdependent group of items forming a unified whole." The following are some of the different types of systems found on dairy farms:

- Housing systems—stanchion barn, tiestall barn, freestall barn, loose housing, drylot, pasture
- Milking systems—pipeline milker, flat-barn parlor, pit parlor
- Feeding systems—pasture, layered feeding, TMR
- Manure handling systems—gutter cleaner, alley scraper, tractor scrape, flush, slotted floor, vacuum loading

- Feed acquisition systems—raise all feed, raise forage and buy grain, buy all feed
- Animal replacement systems—raise heifers, custom raise heifers, buy heifers, buy cows
- **Labor systems**—all family labor, family and hired labor, all hired labor

Producers must select the proper subsystems to support the needs of their dairy. A dairy management system is a group of technologies that complement each other and are regularly found together on farms. The following are a few of the common dairy management systems found in the United States:

- Stanchion barn system—stanchion barn housing, pipeline milking, gutter cleaners, raise all feed, raise heifers, all family labor. A stanchion is a device that fits loosely around a cow's neck and limits forward and backward motion within a stall in a barn.
- Tie-stall barn system—tie-stall barn housing, pipeline milking, gutter cleaners, raise all feed, raise heifers, all family labor. A tie-stall is a type of cow stall in which the cow is constrained by a strap around her neck.
- Grazing system—pit-parlor, pasture feeding, plus various other components.
- Large modern confinement system—freestall barns; flat-barn or pit parlor; facilities to manage manure, feed, and equipment.
- **Drylot system**—drylots (fenced pens or **corrals**) rather than barns, normally pit parlor, plus various other components similar to large confinement systems.
- Robot milker system—freestall housing, robotic milking systems, plus various other components similar to large confinement systems.

Modernization is the process of changing one or more of the components of the dairy's management system. Most existing dairies (in cold climates) now considering modernization are traditional systems (i.e., stanchion or tie-stall barn systems) and are inclined to change to grazing, confinement, or robot systems. Producers in warmer climates additionally have the drylot option. Conversion of these systems can be very costly because the interdependence of the different subsystem types requires many of the subsystems to be changed simultaneously.

The Grazing Option

Profitability should be a major factor in determining which type of system to adopt. Kriegl and Frank (2003) summarized financial data of Wisconsin dairy farms to demonstrate the economic competitiveness of various systems

TABLE 3-2 NFIFO and NFIFO/cwt for Wisconsin dairy systems ($US)

Year	Nonseasonal MIRG	Seasonal MIRG	50–75 Cow Traditional	Large Modern Confinement
NFIFO/cwt 2001	$4.48	$4.78	$2.61	$1.85
NFIFO/cwt 2000	$3.20	$2.68	$1.68	$0.51
NFIFO/cwt 1999	$5.02	$3.97	$2.97	$2.16
NFIFO/cwt 1998	$5.66	$4.32	$3.51	$2.36
NFIFO/cwt 1997	$3.51	$2.27	$1.99	$0.95
NFIFO/cwt 1996	$4.12	$3.62	$3.07	$2.20
Avg.[1] NFIFO/cwt	$4.33	$3.61	$2.64	$1.67
Avg.[1] NFIFO/herd	$49,545	$34,816	$36,039	$172,070
Range in Numbers of Observations	15–22	4–5	196–216	34–50
Range in Average # of Cows/Herd/Year	51–63	46–107	62–63	443–471
5-Year Average[1] # Milk Sold/Cow/Year (1997–2001)	17,125	12,326	19,308	21,648

[1]Simple average of six yearly averages.

under Wisconsin conditions. Table 3-2 shows the average **NFIFO** (**net farm income from operations**) and NFIFO per hundredweight equivalent of milk sold (NFIFO/cwt) of farms using different dairy systems in Wisconsin from 1996 to 2002. They define **MIRG** (**management intensive rotational grazing**) operations as those that harvest up to half of the herds' forage needs via grazing. These herds were classified as seasonal calving or nonseasonal calving. In seasonal calving herds, the dry period of all the cows in the herd overlaps enough to shut down the milking facility for more than a day and preferably for at least a few weeks each calendar year. Nonseasonal herds are those with a calving strategy not meeting the seasonal calving definition. Nongrazing herds were classified as traditional if their average size was under 250 cows, or large modern confinement if their size was over 250 cows.

The average herd size of grazing and traditional herds was similar, and substantially smaller than the large modern confinement group. Profitability based on NFIFO/cwt showed an advantage of grazing systems over traditional and large modern confinement systems. Nonseasonal producers were more profitable by both measures than seasonal grazers. The NFIFO was substantially larger for the large modern confinement group than for any of the other system types.

The Large Modern Confinement Option

In **large modern confinement systems,** cows are housed in large freestall barns and milked in a milking parlor. Such operations are the fastest-growing dairy management system in the United States. Freestall barns offer a controlled environment that minimizes the labor required to feed, bed, and remove the manure associated with the dairy herd. Since the manure is deposited within the barn, this system supports environmentally safe handling of the manure. Cow comfort can be insured by providing comfortable stall bases, easy access to feed and water, and proper ventilation. This type of system can support high production and labor efficiency at a modest investment level. Table 3-2 demonstrates some of the common characteristics of this type of operation: high milk production levels and large profit potential per dairy.

The Drylot Option

The predominant types of cow housing on large dairies in the United States are drylots and freestalls (Smith, Harner, Brouk, Armstrong, Gamroth, Meyer, 2000). The choice of housing is based on climate, management style, and equity available for constructing dairy facilities. Typically, drylot facilities can be constructed where the annual moisture deficit (evaporation rate minus precipitation rate) is greater than 20 inches (Sweeten and Wolfe, 1993). However, frequency and severity of winter rainfall and blizzards are the key selection criteria. These facilities should provide 500 to 700 square feet per lactating cow, depending on the evaporation rate, with 40 square feet of shade per cow. Windbreaks are constructed in areas where winter weather is severe. It is important to realize that drylot housing does not allow managers the luxury of controlling natural risks due to rain, snow, and severe wind chill. The advantage of drylot facilities is a lower capital investment per cow compared to freestall. This type of system is often selected in the semi-arid climates of the southwestern United States, where the climate is very favorable for about eight months of the year. Although temperatures during the remaining four months may be as high as 115°F, with good management practices, production can be kept at comparatively high levels throughout the year.

Drylot System Site Considerations

The following are considerations relating to the proper design of a drylot dairy herd management system (Armstrong, 2004).

Water Supply

An adequate quantity of quality water and a backup water source must be available. Supplies less than 130 gallons per cow-day and salts greater than 1,500 ppm total are considered unacceptable.

Climatic Data

A designer needs to know wind characteristics, rainfall, temperature, and humidity conditions.

Land Area

States may have differing land-area requirements. To prevent groundwater contamination, the following land areas have been recommended for different waste disposal systems of semiarid dairy farms:

Open corral with dry waste	6 acres per 100 cows
Open corral with flush feedline	32 acres per 100 cows
Freestalls with total flush	50 acres per 100 cows

Off-Site Considerations

The site must be within the acceptable travel pattern of the milk processing plant and other dairy services. All-weather roads, utility services, access to feed resources and marketing services, rural housing developments, water management districts, air and water pollution districts, zoning and building codes, and right-of-way restrictions must be evaluated. Legal restraints may close the project and prevent investment recovery, so suitability of site and plans must be checked with appropriate authorities.

Corral Design

The design criteria determine livestock number and space requirements to meet long-range goals. It is important to design for the ultimate goal, not for the first building phase. In warm, semiarid regions housing is simply a fenced pen or corral, with protection from the sun provided by an overhead shade. A feeding slab (manger) with self-locking stanchions is located on one fenceline, and a tank for drinking water is located at an easily accessible site within the corral. The number of lactating cows in a corral should approximate the number of cows that can be milked in the parlor in one hour. Corrals are located to provide the closest possible access to the milking parlor.

Corral space requirements have evolved from dairy practice as the best compromise for manure management, sanitation, and heat stress relief. Table 3-3 represents a range of space recommendations for most semiarid conditions, summarizing the animal grouping arrangement, corral and shade area, and feeder space required for each animal category.

Dry cows are usually provided with a corral of identical size and description as that for lactating cows, or corrals may be split into two. Cows recently dried up are usually located in the more remote pen. The springer pen should conveniently access a maternity pen and treatment facilities. Fresh cows and sick cows should be located in two small pens near the parlor area unless a separate milking facility is used.

TABLE 3-3 Corral, feeder, and shade space recommendations for semi-arid conditions

Animal Class	Per 100 Lactating Cows	Corral Space (ft^2/cow)	Feeder Space (inches/cow)	Shade (ft^2/cow)
Milking cows	100	500	27 or 30	40
Dry cows and springers	15	500	27 or 30	40
Bred heifers (17–26 mos.)	33	350	24	40
Growing heifers (6–16 mos.)	37	300	20	25
Growing cows (6 wks. to 5 mos.)	13	250	17	15
Baby female calves (1 day to 6 wks.)	6	Provide nine individual calf pens		
Maternity or fresh cows	1	500	30	40
Sick cows	2	500	30	40

Replacement animals are grouped according to feeding and management programs. Heifers six weeks to six months of age are grouped in lots of up to 25, with a difference of no more than two months in age or 50 pounds in body weight. Heifers 5 to 15 months and, likewise, breeding-age heifers (14 to 18 months) are grouped in lots of up to 100. Self-locks are convenient for artificial insemination and pregnancy diagnosis. Bred heifers 17 to 28 months are also grouped in lots of up to 100 or can be confined with dry cows and moved to the springer pen about one month before calving.

All corral surfaces should slope 2.5 to 3 percent from the feedline to the cow lane on the opposite side of the corral for good drainage of rainfall. The cow lane then serves as a collector, so it should slope .5 to 1 percent to carry away the collected runoff.

Cattle Shades

Shades should be at least 12 feet high and provide a solid shade pattern. A shade should be located near the center of the corral and oriented with the long dimension in the north–south direction. Corral drainage is enhanced if the shade is perpendicular to the fenceline feed manger.

Placing shades over a feed area will increase feed consumption during the hot summer months and increase milk production. Additional protection from the heat during the hot summer months can be accomplished by placing a spray-line over the feeding area.

Cow Lanes

Since cows must be moved to the barn at least twice daily for milking, the lane and gates should be carefully located. Concrete surfaces 12–16 feet wide are used for cow traffic lanes.

Water Tanks

A float-controlled water tank must be available to provide drinking water at all times. A minimum of one linear foot of tank space should be provided for each five cows. If the water tanks are located adjacent to the cow lane fence, overflow and cleanout drainage from the tank should drain to the surface of the concrete cow lane. Concrete tanks with a roof inhibit heat intake in hot weather and are preferred over steel or fiberglass tanks.

Fenceline Feeding

A flat feed apron lends itself to tractor sweep maintenance. The concrete feed apron and road are sloped about $\frac{1}{8}$ inch per foot to drain away from the feedline. In areas where high winds are frequent, the traditional feed manger may be more satisfactory, especially in the presence of drifting sand.

The cow stand also is sloped for drainage. Lock stanchions are provided along the feed line for all lactating cows and heifers of breeding age. Cattle can be locked in at feeding time and held briefly for breeding, pregnancy diagnosis, or minor treatment.

Treatment Facilities

Treatment facilities are needed for breeding, pregnancy checking, maternity, calf care, vaccination, dehorning, foot care, mastitis treatment, milk fever, and culling. Most routine functions such as breeding, pregnancy checking, and post-calving examinations can be done in lock stanchions at the feedline.

Cows requiring more vigorous examination or treatment by a veterinarian are moved to a special treatment area. A diversion gate at the milking parlor exit lane into a treatment lane is used to separate cows needing treatment. This can be done automatically with computer-controlled gates responding to sensors mounted on the cow's body. A smaller version of this treatment system is also needed in the replacement-heifer area. The head gate and chute system should provide easy access to the cow's mouth, neck, horn area, udder, tail end, side body cavity, and rear muscle area. Side gates may swing in either direction, and provision should be made so that the head gate can be removed and replaced with a hoof-trimming chute.

The treatment lane should provide access to an isolated sand-floor pen equipped with belt lift, floor rings, and overhead supports, and with access for a front-end loader for downed-animal handling. Several small hospital pens and a permanent loading chute should also be available to the main treatment lane.

Calf Housing

Calves can be housed in shaded individual pens with buckets for milk, water, and dry feed or in a permanent calf barn. Portable outdoor calf pens require only a small investment, although labor requirements are high. Sanitary practice includes careful cleaning and relocation of calf pens to an alternate site after each occupancy.

Because of the higher labor requirements of portable calf pens, some producers favor a more permanent calf barn. The most successful model has two rows of individual pens separated by a convenient work alley, all under shade. The pens are made of steel fenceline material set on the well-bedded concrete floor, so that the barn is essentially open on four sides. Canvas or other windbreak systems provide winter protection for the calves.

Feed Storage

Feed storage space requirements for roughage are extensive and important to consider in initial layout design. Feed constitutes about 50 percent of the cost of milk production. It is important to consider potential variations in both short- and long-term requirements. Many dairies in the Southwest purchase nearly all roughage feed at harvest time, so peak roughage storage will approach 65–75 percent of the annual dairy requirement. The fire-insurance carrier usually dictates the maximum haystack size and the distance between stacks. A common stack specification is a 250-ton maximum, with 100 feet minimum spacing between stacks. Haystacks should not be located in front of the corral feed line because the solar heat reflected from the stacks will discourage cattle from feeding during the summer.

When trench silos are used, it is desirable to locate them near the feed mixing center. Two or more silos are preferred to eliminate conflict between filling and feeding functions.

A truck scale and office may be part of the feed service center, but both are usually located near the main entrance to improve security and control. Nonroughage feed supplies are stored in open front bins set on a concrete slab, referred to as a "feed commodity barn."

Water System Design

Peak water use estimated for all animals, and functions per lactating cow per day on a *twice-a-day milking* schedule, are as follows:

- Drinking—65 gallons (including 25 gallons per lactating cow for the replacement herd)
- Parlor and milk room—20 gallons
- Holding-area jet cow wash—25–60 gallons
- Evaporative cooled shades—35 gallons

Well-water pump capacity is selected to deliver peak daily requirements in 12 hours. Since a backup water supply is imperative, an intermediate storage tank supply equal to one day's needs should be installed above ground so that water can flow throughout the system by gravity if necessary.

Waste Disposal

Waste management is relatively simple in a semiarid climate because of the high evaporation rate and the absence of prolonged freezing. The feedline cow platform and area under shade are scraped every few days as needed, and wet manure is dragged to dry areas of the corral. The entire corral area is drag-harrowed lightly to keep manure spread and uniformly exposed to sun and air; this also serves to maintain a uniform surface for good draining of storm runoff. Corrals are cleaned completely once or twice yearly, and the manure is stockpiled locally or spread directly onto farmland.

Required pond storage capacity depends on the number of animals served, wash-down waste (hose wash-down or parlor flush), designed storm runoff reserve volume, nonbiodegradable solids accumulation, evaporation rate, and irrigation interval. Of these, the greatest variable is the wash-down waste. Only 8–10 percent of the total cow manure is deposited in the milk barn and conveyed in the wash water. However, a few dairies have reused their barn water to flush the feedline platform in the corrals. This flush concept involves 50–60 percent of total manure produced. Since sludge buildup is proportional to the manure fraction, pond storage must be enlarged proportionally or the frequency of sludge removal increased.

Any dairy layout design should strive for simplicity and efficiency. Most design requirements are available in published material, but the best sources of innovative ideas are existing dairies—recently built, but with a year or more of operating experience. An engineer can prepare the plans for construction, but the variations involving personal management preferences must come from the dairyman. Working together, the result should be an effective dairy layout design.

The Robot Option

The technology of robotic milking has developed at a rapid pace (Reinemann, 2001). Most of these systems have been developed and were initially installed in Europe (Netherlands, Germany, England, Switzerland, etc.). Canada and the United States began to adopt them later, and now many systems are in use in these countries. These robotic milking systems—or **AMS** (**automated milking systems**), as they are often called—were developed primarily to meet the needs, and the market and social conditions, of single-family owner-operated dairy farms in Europe. A considerable number of similar farms now exist in Wisconsin, Pennsylvania, Minnesota, Michigan, New York, and Vermont.

Moderate-sized farms in the United States account for half of all U.S. dairy farms and roughly 40 percent of all cows. Many of the moderate-sized U.S. farms are located near population centers, where the urban pressures of higher land prices, higher prices for labor, and increasing environmental regulations are significant factors. Health issues, unusual work hours, and onerous working conditions have made obtaining reliable milking labor a major concern of these producers. Robotic milking can provide an option for these farms to reduce the labor requirements of milking and yet continue dairy production.

One of the principal attractions of automatic milking systems is that they may provide an opportunity for producers who want to expand from the 50- to 99-cow herd size to the 100–199 cow range. Producers who want to expand to this level typically cannot justify the cost of a new full-feature parlor and normally must adapt low-cost parlor designs (flat-barn parlors, swing parlors, or renovations of existing barns) or grazing systems that present more modest labor demands on the cropping side of the operation. Robotic milkers, with their capacity of approximately 60 cows per milk station, allow existing producers a growth path that matches their land, location, labor, and management objectives.

Robotic milking, in addition to enhancing economic opportunities for small- to medium-sized farms, may also present a more positive image of milk quality and animal welfare to the public, pose less of an environmental risk, and be generally more socially acceptable, especially near urban areas.

Milking hygiene and milk quality issues are currently the main concerns with robotic milkers. Before their use can become widespread, means must be found to ensure that abnormal milk does not enter the bulk tank, and that the milking machines are cleaned and sanitized.

Although robotic milking may provide some advantages in allowing expansion to a moderate herd size and fostering public acceptance, it is currently a relatively high-cost option. Reinemann (2001) estimated the cost of the equipment and labor for harvesting milk on large farms in the United States to be about $300 per cow per year, compared to $600 to $700 per cow per year for robotic milking. These estimates did not take into account the cost of animal housing, housing of milking facilities, and other differences in producer inputs. Cows milked with a robotic milking system must have constant access to the robot milker, which suggests that freestall housing must be provided. For many existing producers this means that a change to robotic milking requires changing both the milking and housing systems and increases the investment per animal required to make the change.

Factors to Consider When Choosing a System

There are many things to consider when deciding on the type of dairy management system to implement, in addition to the primary considerations of capital cost, profitability, and labor efficiency (Kriegl and Frank, 2003):

- Volume price premiums—many large confinement herds receive a **volume price premium** that can be as large as $1.00 per cwt. (This premium is accounted for in Table 3-2, page 31; it will likely disappear over time.)
- Farm foreclosure—lenders are less likely to foreclose on struggling large-confinement operations than smaller units.
- Herd turnover—recent Wisconsin **DHIA** (**Dairy Herd Improvement Association**) data indicate little difference in total turnover

rates between herd sizes, but some observers feel that the ability to do involuntary culling is more difficult in large herds.

- Labor availability—most Wisconsin grazing and traditional dairies rely primarily on the labor and management of one family. This commitment may not be acceptable to some families. Larger operations require multiple workers and therefore allow for flexibility in work scheduling, as well as specialization, which minimizes employee training.
- Operations management stress—graziers "employ" their cows to do much of the physical labor in feed harvesting and manure spreading. The traditional and large-confinement systems rely much more on machinery and human labor for these tasks. Many farm operators who have had several years of experience operating a traditional system before switching to and gaining experience operating a grazing system of a similar size, say that they used about the same number of hours in each system. However, they also say that their work is less physically intense, they experience less stress, and they get more satisfaction from providing the labor and management in their grazing system. Owners of large confinement systems often do most of the management and a small percentage of the labor. This may lead to a range of stress levels for the owner. A manager of a highly successful large confinement system may experience far less stress and have far more free time than the owner of a traditional or a grazing system. The owner of a struggling large confinement farm may be as tied to the farm and under far more stress than the average person operating a dairy system that relies primarily on owner labor.
- Land use conflicts—many large-confinement farms have met considerable public resistance from nonfarm rural communities, traditional farmers, and environmental groups.
- Economic flexibility—a well-planned grazing operation will likely be able to recover a larger percentage of its investment if the decision is made to quit farming or change systems.
- Herd start-up size—grazing can be done with or without practices and technologies such as seasonal calving, milking parlors, TMR, and so on; therefore, producers can start with a smaller herd. Large-confinement herds must be substantial to justify the capital expenditures needed.
- Herd production level—three factors influence the profitability of a business: income generation, operating expense control, and investment control. Production levels of grazing herds are often lower than those of traditional herds or modern large-confinement herds. Graziers tend to emphasize operation cost and investment control. Traditional Wisconsin producers tend to emphasize income generation. Although many graziers are financially

competitive at production levels lower than other systems, they may be even more competitive if they do not sacrifice production, since cost and investment savings are not automatically created when production is reduced. Dairies transitioning between systems may not be able to afford much of a production decline. Whichever system is selected, money should be spent to increase profitability by optimizing the three profit factors listed previously.

What Size and Type of Milking Facility Will Be Used?

To determine the site size needed to support your long-term plans, you need to define the type, size, and expected throughput of the milking parlor being considered. The selection of the size and type of parlor, the number of milkers to be used, and the milking procedures deployed will determine its capacity. Normally, parlors are planned to fully utilize the efforts of a certain number of milkers. Industry standards indicate that one milker can manage from a double-8 to a double-12 parallel or herringbone parlor, whereas a double-14 to double-24 is considered a two-person parlor. Milking shifts of 6.5 hours (for 3X milking) and 10.5 hours (for 2X milking) are used to determine the capacity of a parlor and allow time for setup, cleanup, and maintenance of the milking facility.

The throughput of a milking parlor is normally defined in terms of the average number of cows milked per hour (cows/hour), or the number of times the parlor will be filled per hour (turns/hour). The throughput of the milking parlor determines the optimum pen size and maximum herd size for the operation. Current recommendations suggest that pens be sized to allow a group of cows to be milked in 60 minutes or less if milked two times per day (2X), 45 minutes or less if milked three times per day (3X), and 30 minutes or less if milked four times per day (4X). This rule ensures that cows have sufficient time to eat and lie down, with a reasonable amount of time standing in a holding area, away from feed and water, waiting to be milked. For example, if a system is designed with a parlor that is expected to have a milking capacity of 72 cows/hour, then pens should be designed to hold 72 cows for 2X milking or 54 cows for 3X milking. For planning purposes, you should normally assume that herds are composed of 16 percent dry cows (12 percent far-off and 12 percent close-up) and 84 percent milking cows (of which 4 percent are considered just fresh). These values may vary by herd and over time, depending on calving distribution, dry-period length, and dry-cow management of the herd.

Table 3-4 shows the maximum herd size to expect based on the parlor selected. Using this approach, a producer who plans to build a double-12 parlor, milking at four turns per hour, two times per day (2X), would need to reserve space for a 1,200 cow facility, with pens that hold 96 cows each.

TABLE 3-4 Herd size potential based on the capacity of different parlor sizes

	D-8	D-12	D-16	D-24	D-36
Number milkers	1	1	2	2	3
Stalls per side	8	12	16	24	36
Total stalls	16	24	32	48	72
Expected turns/hr	4.5	4	4.5	4	4.5
Expected cows/hr	72	96	144	192	324
Total milking cows if 6.5-hr shift, 3X	468	624	936	1,248	2,106
Total herd with 16% dry	557	743	1,114	1,485	2,507
Total milking cows if 10.5-hr shift, 2X	756	1,008	1,512	2,016	3,402
Total herd with 16% dry	900	1,200	1,800	2,400	4,050

How Will the Parlor Complex and Freestall Barns Be Arranged?

In the past, there have been two common ways to arrange freestall barns and parlor/holding area complexes. The **H-style** configuration has two freestall barns with a parlor/holding area complex between the two barns (Figure 3-1). The **T-style** configuration has the parlor/holding area complex perpendicular to and attached to one of the freestall barns, with the second freestall barn behind and parallel to the first barn (Figure 3-2). Lately a third complex type, referred to as the **modified H-style,** has become very popular. It has the parlor complex arranged parallel to both barns like the H-style, but the parlor complex is located beside the barns rather than between them (Figure 3-3). The ends of each barn shown are labeled as a pen, but center drive-through barns, or barns with pens located behind other pens, may contain two or more pens at each end.

The advantages of the H-style arrangement are as follows.

1. Space is available behind the parlor/holding area for the special-needs barn; this is a convenient location for sick and fresh cows. Pens in this barn are closest to the parlor, so these cows have less distance to walk; it is centrally located to both freestall barns, so employees can easily monitor cows, and work activities are near the office and parlor areas.
2. The orientation of the holding area is the same as in the freestall barns, providing proper natural ventilation to both.

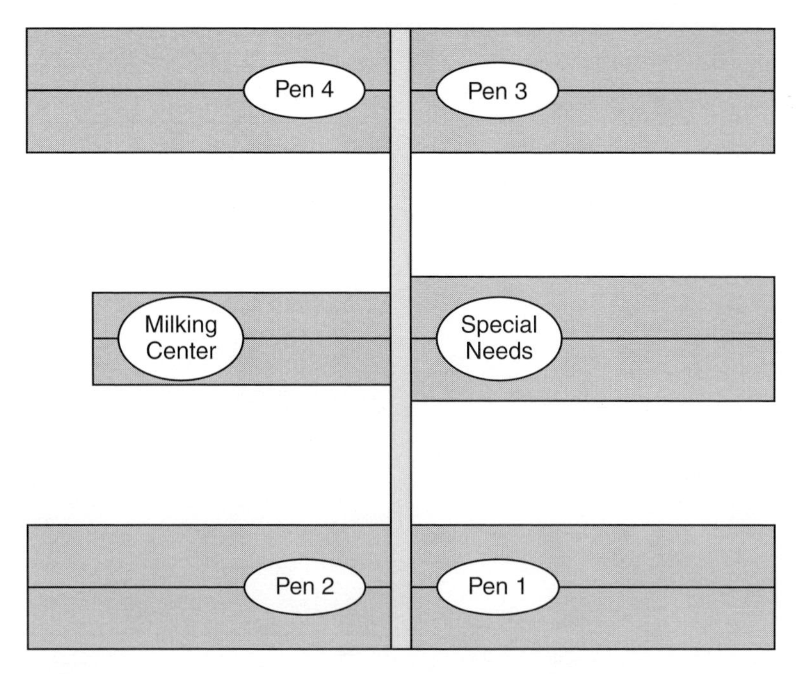

FIGURE 3-1 H-style barn complex: barns with end-to-end pens parallel to the milking center, with special-needs barn behind the milking center, and one barn complex on each side of the milking center.

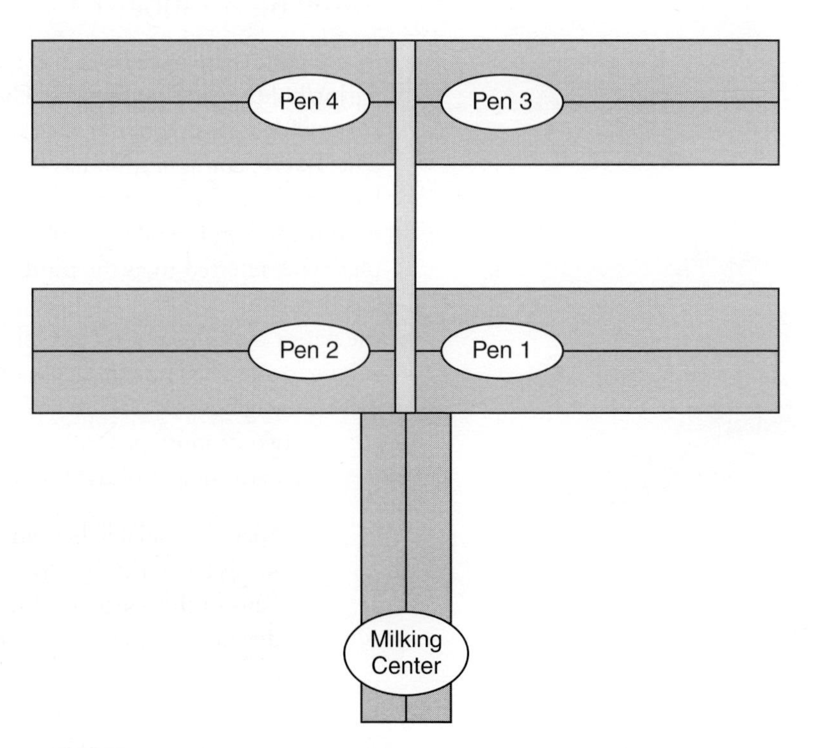

FIGURE 3-2 T-style barn complex: barns with end-to-end pens perpendicular to the milking center (no special-needs barn shown, but pen 1 would often be used for this purpose).

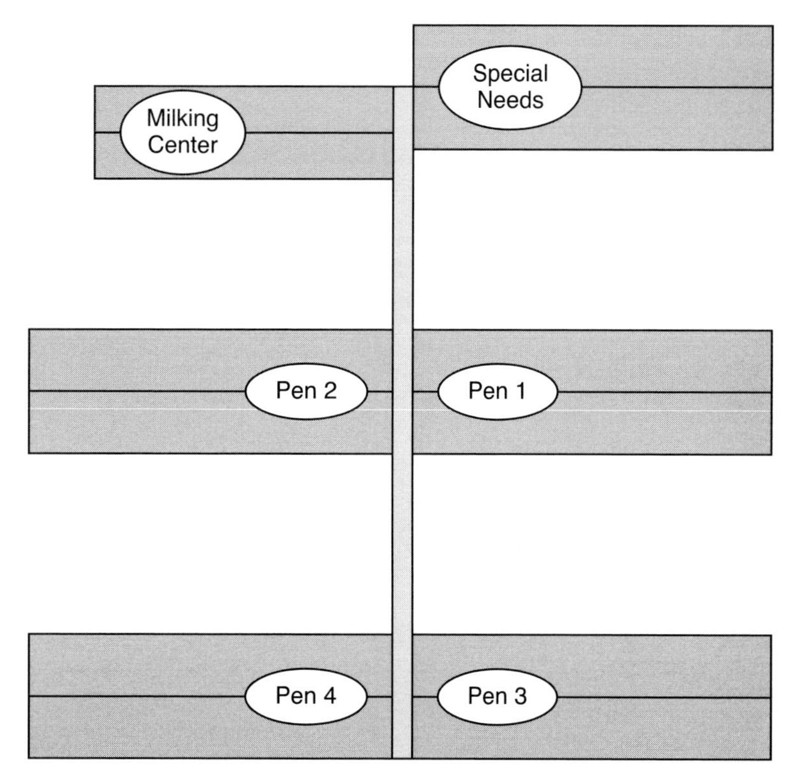

FIGURE 3-3 Modified H-style barn complex: barns with end-to-end pens parallel to the milking center, with special-needs barn behind the milking center, and both barn complexes on the same side of the milking center.

3. Cow flow from either freestall barn does not interfere with the other freestall barn, which allows more time for nonmilking activities (feeding, bedding, etc.) to be performed between milkings for each barn.
4. The average walking distance for groups of cows often is shorter.

The advantages of the T-style arrangement are as follows.

1. Manure from the holding area can be easily incorporated into the manure handling of the attached freestall barn.
2. Only one connector barn or alley is needed.

The advantages of the modified H-style arrangement are as follows.

1. Space is available behind the parlor or holding area for the special-needs barn; this is a convenient location for sick and fresh cows. Pens in this barn are closest to the parlor, so these cows have less distance to walk. Work activities are near the office and parlor areas.
2. The orientation of the holding area is the same as in the freestall barns, providing proper natural ventilation to both.

3. Since the special-needs barn is offset from the parlor or holding area complex, the drive-through feed alley for the special-needs barn can be easily accessed.
4. Since all cows must enter and leave the parlor complex via the same connector barn, it allows one site to access all cows going to or returning from the parlor. This can decrease the number and implementation cost of palpation rails, sort gates, animal oilers, and so on.

If the T-style configuration is selected, some portion of the two freestall barns may be used to house special-needs animals, a special-needs barn can be attached to the parlor building, or a separate barn can be constructed near the parlor or holding area or freestall barn complex. When choosing among these alternatives, labor efficiency and operator convenience should be considered. The ability to easily monitor, move, and treat animals with special needs is very important.

How Will Manure Be Handled?

Manure management of freestall barns includes manure collection, removal, and storage, which together constitute the **manure handling system.** Most modern dairies collect manure from manure alleys with a tractor scraper or automatic alley scraper or a water-flush, slotted-floor, or vacuum-loading system. Manure is often collected in the middle or end of the barn and moved to a reception pit or outside permanent storage using a concrete cross channel or a round-pipe flume system. Manure collected in reception pits can be hauled, pumped, or gravity-flowed to clay-, synthetic-, or concrete-lined storage units. Slotted-floor systems store manure below the floor in concrete-lined tanks or use gravity channels to channel the manure to outside storage areas. Vacuum-loading systems, where manure is removed from manure alleys and hauled to storage, are relatively new but have advantages that should lead to their increased use. Each of these systems will be described in more detail in Chapter 11, "Manure Handling Options." At this point, it is important to know which manure handling system will be used so the facility can be designed to accommodate the features selected, as in the following examples:

- If alley scrapers are used in a cold climate, additional space should be allocated to keep drive units and wheels inside the building to minimize maintenance concerns during inclement weather.
- If alley scrapers are used and manure is taken to the center of the barn, the slope of the barn floor should be minimized (0 to .5 percent) to prevent scraper lift-up.
- If alley scrapers are used, the facility should be designed to avoid their use in pens of sick and recently fresh cows, because of

concerns that the alley scraper may not stop and will injure a cow that lies in the manure alley.

- If flush manure removal is chosen, the barn floor should have sufficient slope (1 to 3 percent) to allow manure to be removed, and freestall bases may be cantilevered to remove manure near the freestall curb.
- In cold climates, removal and storage of frozen manure must be anticipated, and doors and manure stacking pads considered.
- The primary manure handling system should include an alternative plan that can be easily implemented when the primary system is not functional.

How Will Cows Be Fed?

It is important at this point in the planning process to know generally what feeds will be given, what amount of each feedstuff will be grown, and how the feed will be stored, mixed, delivered, and fed. For most herds a total mixed ration (TMR) will be delivered to animals along a fenceline feed apron using a portable TMR mixer. This is a labor-efficient method of feeding different rations to groups of animals at various locations on the farmstead. Wet forage will normally be stored in upright silos, plastic bags, bunkers, or piles, and dry hay will be baled. Grain and concentrates can be stored in upright bins, which minimize wastage and feeding inaccuracies, or in an open-fronted commodity shed. The implications of each of the feed storage and feed management choices should be considered when sizing facilities, defining site-size requirements, and developing facility layouts that support efficient feed delivery routes. (This will be discussed in more detail in Chapter 10, "Feeding the Dairy Herd.")

What Freestall Bedding Material Will Be Used?

To maximize milk production, dairy cows must have clean, comfortable beds, and a clean environment in which to eat, drink, and move. Many different freestall types are used to provide a comfortable bed. Chapter 7, "Freestall Design and Bedding Materials," will define the characteristics of different freestall types. The important thing at this time is to determine if sand, lime, or any other bedding type that is known to settle out of the manure-bedding combination will be used. For example, sand bedding entails the following consequences:

- Wide concrete cross-channel manure transfer systems should be avoided because of sand settling concerns.
- Flume-type manure transfer systems that utilize a buried pipe to transport manure may be avoided because of plugging concerns.

- Alley-scraper manure removal systems may be avoided because of excessive wear expected to blades and turn wheels.
- Slotted-floor systems should be avoided because of sand settling and removal concerns.
- Manure storage may need to be placed closer to animal housing to support the manure removal and transfer system.

How and Where Will Animals Be Handled?

All animals must periodically be isolated and restrained for physical examination, vaccination, artificial insemination, pregnancy check, treatment, dehorning, calving, and so on. The animal flow—the paths followed by the cow as she is moved to the parlor, to her home pen, or to a different location for treatment—and any location changes needed throughout the lactation should be considered. Facility design and equipment selection influence work routines, labor requirements, and animal stress levels associated with each of these activities.

The decision of how and where these tasks will be performed influences the overall design and the resulting labor efficiency of the facility. Here are a few examples:

- If self-locking manger stalls are used, three- and six-row barns may be avoided because of feed intake concerns and the inability to capture one-third of the cows at once.
- If automatic sort gates are used, parlor designs that have a single return lane may be chosen to decrease the number of sort gates, palpation stations, catch pens, supply storage areas, and labor requirements.
- The need to have a conveniently located maternity and treatment area may lead to choosing the H-style barn configuration over the T-style configuration, so adequate building space can be provided near the milking facility.

An animal handling system affects management, facility layout, work routines, and labor requirements. Any additional initial costs should be prorated and added to the ongoing labor requirement to arrive at an estimated annual cost of using each system. Putting a value on daily convenience is sometimes difficult but has substantial impact when considering the implications of a structure that will last 10–20 years.

Which Parts of the Existing Facilities Will Be Used, and for What Purposes?

An honest evaluation of existing facilities should be conducted to determine their usefulness, but care should be exercised to prevent overvaluing them. Situating new facilities near an old one should be considered if doing so does not prevent long-term growth of the operation. The effects of using existing facilities should be carefully evaluated:

- If dry cows are to be housed at a different site, how frequently do they need to be hauled, and what are the associated costs?
- Milking sick and fresh cows in a second pot-barn parlor can help prevent contamination of salable milk.
- If sick or fresh cows are to be milked at a separate location, what is the effect on labor, equipment, cleaning supplies, and so on?
- Using an old facility to freshen cows is discouraged for small operations (fewer than 500–1,000 cows) because it requires a second crew, and supervision of these animals is often insufficient to avoid management problems.
- If feed storage is to be added near existing feed storage units, does the site support the long-term needs of the operation? How far will feed need to be hauled, and at what cost?

SUMMARY

The facility planning process is critically important to the dairy producer because design decisions affect the efficiency and safety of animals and workers over the long term. It is highly recommended that planners spend time visiting other dairies, talking to experts, and reading current literature to ensure that what is planned is the best choice, considering the goals and management style of the owners. When observing other dairies, be sure to look at more than the stainless steel and concrete. Try to arrange farm visits when animal-handling activities are being performed and when the owner has time to answer your questions. Try to understand how the facility is being used, not just what it looks like. Watch how each type of cow is being handled, the labor required, and the type of person needed to perform each role. Answering the seven questions reviewed in this chapter will set the stage for the development of a facility design that supports your objectives.

CHAPTER REVIEW

1. List the four financial aspects of a business that determine its borrowing capacity.
2. Which system (milking, housing, manure handling, feeding, feed storage, or animal handling) most affects the overall design of a dairy facility?
3. Explain how the design of a housing system impacts milking and manure-handling systems.
4. List the four primary characteristics of a confinement dairy management system, and the four primary characteristics of a drylot dairy management system.
5. Consider how climate affects a dairy's functioning and design. How do heat and cold weather affect housing and feed acquisition systems? Compare the modernization options and concerns that a dairy in the West has, compared to an operation in the Midwest, and vice versa.

REFERENCES

Armstrong, D. V. (2004). *Dairy design for a semi-arid climate.* [Technical paper.] Tucson: University of Arizona.

Bickert, W. G. (1998). *Sorting, handling and restraining lactating cows for treatment and other purposes,* (pp. 44–50). St. Joseph, MI: American Society of Agricultural Engineers.

Bickert, W. G., Holmes, B., Janni, K., Kammel, D., Stowell, R., & Zulovich, J. (2000). *Midwest Plan Service dairy freestall housing and equipment handbook* (7th ed.; MWPS-7). Ames: Iowa State University.

Graves, R. E. (n.d.). *Dairy cattle handling and behavior.* [Handout based on excerpts from the *Dairy reference manual.*] Ithaca, NY: Natural Resource, Agriculture, and Engineering Service.

Kriegl, T., & Frank, G. (2003). *An economic comparison of Wisconsin dairy systems.* Madison: University of Wisconsin Center for Dairy Profitability.

Martin, J. C., III. (1998). *Siting large dairy facilities,* (pp. 29–36). St. Joseph, MI: American Society of Agricultural Engineers.

Pajor, E. A., Rushen, J., & de Passille, A. M. (2000). Cow comfort, fear and productivity. In *Proceedings from the Dairy Housing and Equipment Systems: Managing and Planning for Profitability Conference,* (pp. 24–37). Camp Hill, PA, February 1–3. Ithaca, NY: Natural Resource, Agriculture, and Engineering Service.

Reinemann, D. J. (2001, March 20). *Evolution of automated milking in the USA.* Handout presented at the First North American Conference on Robotic Milking, Toronto, Canada.

Smith, J. F., Harner, J. P., Brouk, M. J., Armstrong, D. V., Gamroth, M. J., & Meyer, M. J. (2000). *Relocating and expansion planning for dairy producers.* Manhattan: Kansas State University. (MF2424, January.)

Sweeten, J. M., & Wolfe, M. L. (1993). *Manure and wastewater management systems for open lot dairy operations.* In *Proceeding of western large herd management conference,* Las Vegas, NV. Manhattan: Kansas State University Agricultural Experiment Station and Cooperative Extension Service.

Chapter 4

Animal Handling Needs

KEY TERMS

animal handling activities
automatic sort gate
bST injection
catch lane
catch pen
home-based animal handling system
palpation rail system
people-pass
self-lock
special-needs barn
treatment-area-based animal handling system

OBJECTIVES

After completing the study of this chapter, you should be able to

- identify the key components of home-based and treatment-area-based animal handling systems.
- understand how each system affects overall facility operation.
- understand how each system affects overall facility design.
- evaluate an existing operation to determine which animal handling system is ideal, given the operation's existing facilities and short- and long-term goals.
- develop a facility design that works well with modernization phasing plans.

Methods and Locations of Animal Handling

All animals must periodically be isolated and restrained for physical examinations, vaccinations, artificial inseminations, pregnancy checks, treatment, dehorning, calving, and so on, on a regular basis. The animal flow consists of the paths followed by a cow as she is moved to the parlor, to her home pen, to a different location for treatment, and so on. This flow, and any location changes needed throughout lactation, should be considered when choosing facility design and equipment. These choices influence work routines, labor requirements, and animal stress levels associated with each of the aforementioned activities.

Cattle's fear of people can be a major source of stress. Stressed animals produce less milk, and milking efficiency is reduced. Stressed cattle are also difficult to handle and increase the risk of accidents and injuries for handlers and animals. Poorly designed handling facilities cause animals to balk. Properly designed handling facilities lead to easier flow of animals, reduce the need for rough handling, and result in tamer, less fearful animals. When planning treatment facilities, the following recommendations should be considered.

- One person should be able to isolate and restrain an animal safely and conveniently.
- Components should be selected and constructed to reduce the possibility of injury to operators and animals.
- Construction should be designed to withstand the abuse given by 1,500-pound cows and by equipment used to clean the area.
- Access to running water, medical supplies, records, and parking for the veterinarian's and hoof trimmer's service vehicles should be included.
- Good lighting should be provided.
- Detailed drawings that show how gates will be used to form a funnel to direct reluctant animals into a stanchion or lockup, and how people will access restrained animals when performing needed activities, should be thoroughly reviewed.
- Animal handling systems designers should consider system impact on left- and right-handed veterinarians or managers.
- A heated storage area for supplies and equipment should be provided near the animal treatment area.
- **People-passes** and bypass lanes should be provided near animal treatment areas to support animal movement.
- If animals will be calved on-site, the maternity location should be situated so that animals may be observed easily and often.
- Sick cows should be housed separately from the maternity area.
- If dry cows are to be housed at a remote location, the facilities should be designed to support their separation and movements.

Animal Handling—Possible Systems

Most new parlor or freestall operations use one of two types of animal handling systems. The **animal handling activities** of sorting, restraining, and treating are often done in the freestall unit where the animals are housed (home based), or in some special area away from where they are normally housed (treatment-area based).

Home-based animal handling systems normally utilize **self-locks,** also known as self-locking manger stalls (Figure 4-1), where cows lock themselves in place upon returning to a manger full of fresh feed after being milked. The self-locking feature is activated when the animal puts her head in a stanchion to eat. A few dairy producers treat animals by cornering them in a freestall. This practice is discouraged because of the safety concerns if the animal moves or the worker slips.

Treatment-area-based handling systems use sort gates to separate selected animals from their group as they leave the milking parlor. These can be standard sort gates manually controlled by the parlor operator, or **automatic sort gates** controlled by a computer if animals are electronically identified. Animals sorted in this manner may be directed into a **palpation rail system** (also referred to as a management rail system) or placed in a holding pen and handled using a head chute or some other restraint system.

Home-Based Animal Handling Systems

Dairy managers who select the home-based system must evaluate the cost of the self-locking manger stalls (about $60 per head lock) versus the cost of a

FIGURE 4-1 Self-locking manger stalls are an efficient means to restrain, sort, and treat animals.

separate treatment area, plus any labor savings over time. Producers report the following advantages of the self-locking manger stalls.

- Handling of cows is less traumatic since they are treated in familiar surroundings.
- Cows may eat their proper ration while waiting to be treated.
- No time is wasted returning animals to their lot after treatment, because they are restrained in their own pen.
- Manger uprights minimize the effect of boss cows dominating a large section of the feed bunk, and can decrease feed wastage.
- Large numbers of cows can be automatically restrained, saving labor for routine tasks such as tail chalking, hoof spraying, **bST injections,** and so on.
- Manure from restrained animals is handled with normal procedures and can be incorporated into the planned manure handling system.
- Locking cows after milking allows teat sphincter muscles to close before the cow lies down, thereby decreasing the likelihood of mastitis.
- Parlor efficiency can be improved: the flow of cows leaving the parlor need not be channeled through a narrow sort lane, and no operator time is spent to sort or move animals.

Users of self-locking manger stalls sometimes express concerns about the extra noise generated by some brands, and the difficulty of finding a specific animal when the cows are caught and restrained in a random order.

Some producers have installed self-locks in only a portion of each housing area, which can be gated off and used to treat groups of animals moved from a sort area. This technique can decrease the initial expenditure for self-locks but may complicate animal management, since animals will probably show a preference for eating in the section containing no self-locks.

Treatment-Area-Based Animal Handling Systems

In treatment-area-based systems, animals are sorted and taken to a special place for restraint and treatment; with this system, the manager must be concerned with the length of time the animal will be away from her home pen and how she will be returned. Labor requirements, availability of feed and water, the effects of the additional stress placed on animals, and handling of manure are some of the issues to consider when making this choice.

With treatment-area-based systems, cows are often sorted as they leave the milking parlor. Cows must be diverted through a narrow alley that allows them to be identified and diverted to a **catch lane, catch pen,** or palpation rail. This animal selection process can be installed anywhere in the path as the animal returns home. Manual sorting should be done near the rear of the parlor to be easily viewed by the operator, but automatic sorting should be located near the end of the return lane to improve cow movement. Maintaining a smooth flow of animals, walking slowly, generally will improve an automatic sort gate's ability to accurately select desired animals. Animals that bolt through a sort gate can be incorrectly selected or missed, depending on the position of the sort gate.

Once cows are sorted, they can be restrained and treated using a palpation rail or a chute located in the catch lane, or may be taken to a catch pen containing self-locking stalls. One major consideration with treatment-area-based systems is that animals returning to their home pen after being treated may use the same traffic lanes as animals being milked. This can cause delays and additional labor to move gates, and so on, to prevent mixing of groups.

Palpation Rail Systems

Palpation rails are simple structures that allow a group of animals to be restrained in a common area (Figure 4-2). Front pipes position animal heads, and a rail in the rear restrains the animals but is low enough to allow animals to be palpated. Animals are positioned and restrained in a herringbone fashion as they are given rectal examinations, are bred, or are given shots.

FIGURE 4-2 Palpation rails are an inexpensive way to restrain and treat groups of animals. They should be located near normal animal traffic areas, but far enough away from the milking parlor to avoid animal backup.

An advantage of palpation rail systems is that they can be placed in normal cow flow areas, simplifying the selection and restraining of animals. This can also be a disadvantage, in that animals must be treated promptly to provide space for subsequent animals, or a buildup of animals waiting to be treated can disrupt animal flow. In addition, the veterinarian or herdsperson may need to be on-site for an extended time, waiting for all animals to be selected. Palpation rails are often considered as an alternative to self-locking manger stalls because, for large herds, the cost per animal is less, and they offer safer working conditions since they prevent animals from kicking workers behind them.

Separate Treatment Areas

As herd size increases, the need increases for a special area where cows can be taken and treated (Figure 4-3). This treatment area should be conveniently located, have a head chute and medical supplies, and be well lighted. This special treatment area is often located near the maternity and sick-cow areas so that facilities and equipment can be shared, and workers who are engaged in other activities can monitor cows. When these areas are combined, they are often referred to as **special-needs barns.**

One or more catch pens should be designed to support the needs of the dairy. If animals are selected from several different milk groups and

FIGURE 4-3 Dairy operations should have a conveniently located, well-lighted area to handle cows needing special attention. Notice the people-pass to the right of the catch chute, which is large enough to allow a worker, but not a cow, to pass through the fence.

FIGURE 4-4 Sort gates are often used to select cows needing special attention.

placed in a common pen, they will need to be sorted later, which increases labor inputs. If animals are to remain in these pens for a substantial amount of time waiting to be treated, feed and water should be provided.

This special treatment area can be smaller for herds handled mostly in palpation rails or self-locks, since it will be used mainly for major surgeries, hoof trimming, and so on. For herds that do not use one of these animal-handling systems, this area should be larger, and keeping animals from different groups separated should be considered.

Electronic Sort Gates

For many large herds, electronic sort gates are used that automatically identify and sort animals that need special attention (Figure 4-4). Animals are identified by an electronic transponder that is temporarily or permanently placed on the animal.

Temporary identification is normally attached to the leg of the animal during milking and causes the animal to be selected as she returns from the parlor. This is a less expensive system but requires extra labor in the parlor, so it may slow the milking process, and it is governed by the milker's ability to identify cows needing attention.

The permanent identification system is more expensive but does not require milker intervention. Since this system is computer driven, a highly accurate and up-to-date database is imperative for proper functioning. Users thinking about this alternative should consider the costs of equipment and the labor to place and replace neck bands on cows; they must also consider how mistakenly sorted cows will be handled.

Special-Needs Area Location

In Chapter 3 you read about three common ways to arrange freestall barns and parlor/holding area complexes. The H-style configuration has two freestall barns, with a parlor/holding area complex between the two barns. The T-style configuration has the parlor/holding area complex perpendicular to and attached to one of the freestall barns, with the second freestall barn behind and parallel to the first barn. The modified H-style configuration has the parlor complex arranged parallel to both barns like the H-style, but the parlor complex is located beside the barns rather than between them.

A major factor in deciding between these designs is the location of the special-needs area. With the H-style and modified H-style, the special-needs area is located behind the parlor or holding area complex. If the T-style configuration is selected, some portion of the two freestall barns may be used to house special-needs animals; a special-needs barn can be attached to the parlor building; or a separate barn can be constructed near the parlor or holding area or freestall barn complex. When choosing among these alternatives consider labor efficiency and operator convenience. The ability to easily monitor, move, and treat animals with special needs is very important.

As stated previously, one of the main issues in the overall design of a new dairy facility is how animals will be sorted, handled, and restrained for treatment. The decisions concerning how and where these tasks will be performed influences the overall design and the resulting labor efficiency of the facility; for example:

- if self-locking manger stalls are used, three- and six-row barns may be avoided because of feed intake concerns and the inability to capture one-third of the cows at once.
- if automatic sort gates or palpation stations are to be used, parlor designs that have a single return lane may be chosen to decrease the number of sort gates, palpation stations, catch pens, supply storage areas, and labor requirements.

The need to have a conveniently located maternity and treatment area may lead to choosing the H-style barn configuration over the T-style configuration, so that adequate building space can be provided near the milking facility.

SUMMARY

Knowing how animal management activities will be performed is very important when designing a parlor or freestall complex. Parlor complexes designed with return lanes on each side of the holding area work well with home-based systems but not with treatment-area-based systems, because of the need for two sort gates and two catch areas. If a treatment-area-based system is being designed, a parlor complex that allows all animals to return from the milking parlor in a single return lane should be considered. It is important to remember that both home-based and treatment-area-based systems will work, but the effects on management, facility layout, work routines, and labor requirements should be considered. Any additional initial costs should be prorated and added to the ongoing labor requirement to arrive at an estimated annual cost of using each system. Putting a value on the daily convenience is sometimes difficult but has substantial impact when one considers the implications of a structure that will last 10–20 years.

CHAPTER REVIEW

1. What are the two primary causes of stress in dairy cattle?
2. What are the three common results of dairy cow stress?
3. List the advantages and disadvantages of home-based animal handling systems.
4. List the advantages and disadvantages of treatment-area-based animal handling systems.
5. Draw H-style, modified H-style, and T-style barn configurations on a sheet of paper, being sure to include both the freestall barn and the parlor or holding area in your diagrams. In each of these configurations, identify (or add, if necessary) an area to house special-needs animals. Using arrows, diagram how animals will flow through the site. What animal sorting problems might be encountered? What are your solutions to these problems? Are your solutions affordable, given standard dairy operational styles and available labor?

REFERENCES

Bickert, W. G. (1998). Sorting, handling and restraining lactating cows for treatment and other purposes. In *Fourth International Dairy Housing Conference* (pp. 44–50), St. Louis, MO, January 28–30. St. Joseph, MI: American Society of Agricultural Engineers.

Bickert, W. G., Holmes, B., Janni, K., Kammel, D., Stowell, R., & Zulovich, J. *Midwest Plan Service dairy freestall housing and equipment handbook* (7th ed.; MWPS-7). Ames: Iowa State University.

Graves, R. E., (n.d.). *Dairy cattle handling and behavior.* [Handout based on excerpts from the *Dairy reference manual.*] Ithaca, NY: Natural Resource, Agriculture, and Engineering Service.

Pajor, E. A., Rushen, J., & de Passille, A. M. (2000). Cow comfort, fear and productivity. In *Proceedings from the Dairy Housing and Equipment Systems: Managing and Planning for Profitability Conference,* (pp. 24–37), Camp Hill, PA, February 1–3. Ithaca, NY: Natural Resource, Agriculture, and Engineering Service.

Chapter 5

Animal Housing Options

KEY TERMS

annual cost
average annual total cost
back pen
bedded pack
drive-by feeding
drive-through feeding
drover lane
early fresh cow
freshen
front pen
lifts
outside feeding
post-calving
retrofit
roof pitch (4:12)
special-needs animal
transition cow

Objectives

After completing the study of this chapter, you should be able to

- identify the physical differences between 2-, 3-, 4- and 6-row freestall barn designs.
- understand the advantages and disadvantages of each type of design.
- understand how each type of design affects facility operation and maintenance.
- decide which design will work best with an existing dairy's modernization plan.

Chapter 4 covered the need for and characteristics of facilities that allow the proper handling of cows with special needs. These **special-needs animals** are normally housed in a central location, where management can monitor them and support their needs. The remainder of the herd is often housed in separate barns that are conveniently located to allow cows to travel to and from the milking facility with minimal human effort or animal trauma. This chapter presents a discussion of the features and benefits of different freestall barn designs currently in use by dairy producers. Selecting the freestall barn design that complements your animal-handling protocols and facilities is very important. The barn type selected will define the dairy's space requirements. (Proper site selection will be the subject of Chapter 8.)

Animal Grouping Strategies

Modern dairy herds are normally large enough to allow the operator to house together animals with like needs and to manage them as groups. These animal groupings can support the nutritional, reproductive, or health needs of the animals and the desired management practices. The number of different groups maintained by a dairy is often based on herd size and the capacity of the milking system selected. The first rule of pen-size determination is that the largest group of cows should be milked in one hour or less for 2× milking, or 45 minutes or less for 3× milking. Using this rule, barns are designed with enough pens to support the number of cows desired or that can be milked with the existing or planned milking facility. In new facilities, animal housing should be designed to support current desired grouping and management procedures but be flexible enough to allow grouping changes that can support new technologies and procedures that will be introduced in the future. When facilities are being planned, pens for milking cows, dry cows, and those with special needs should be designed to support expected numbers of each animal type. Table 5-1 gives an example of the housing needs for a 1,000-cow dairy, based on the expected number of each animal type. The percentages shown are based on expected size of the total herd, which includes both milking and dry cows. The table shows the relative number of each type of animal that could be expected in a 1,000-head dairy if a perfectly even calving distribution could be maintained. Since animals are constantly entering and leaving the herd, housing for more than 1,000 animals is needed to accommodate heifers waiting to **freshen.** But since calving is not always uniform, the number of individual calving pens shown should be considered a minimum. Housing for the **post-calving** and **early fresh cows** should also be designed to comfortably accommodate additional animals if an uneven calving is experienced. Sizing pens for these groups for a 90 percent stocking rate is one way to accomplish this.

A large modern dairy facility normally is composed of multiple barns. These barns normally are located far enough apart (approximately 100 feet) that they do not interfere with the natural ventilation of other barns and

TABLE 5-1 Normal percentages of each type of animal in a dairy herd, and expected housing needs of a 1,000-cow herd.

Type of Animal	Percent of Cows	Number of Cows
Cows with salable milk:		
Healthy cows	76%	760
Slow and lame	2%	20
Early fresh	3%	30
Total salable milk	81%	810
Cows with unsalable milk		
Sick cows	2%	20
Post-calving cows (4 days)	1%	10
Post-calving heifers (1 day)	.3%	3
Maternity (1 day)	.3%	3
Total unsalable milk	3.6%	36
Dry cows		
Far-off	12%	120
Pre-calving dry cows	4%	40
Pre-calving heifers	4%	40
Total dry and close-up	20%	200
Total housing needed	104.6%	1,046

close enough to minimize walking distance from housing to milking facilities. Normally the herd is considered to be composed of three types of animals: healthy milking cows, dry cows, and cows with special needs. Healthy milking cows are normally housed in large barns containing several pens sized to match the parlor's milking capacity. Dry cows often are housed in separate facilities away from the milking herd. Cows with special needs are often placed in a barn (often referred to as a special-needs barn) close to the parlor, where they can be monitored and treated. Herd size, construction costs, and the management approaches planned for the dairy influence the type and location of each barn and the types of animals that will be housed in each. Table 5-1 shows the approximate percentage of each type of animal expected. These values may vary by herd and time of year because of seasonal calving differences and so on. Facilities for large herds should be designed to house each of these groups separately and have enough flexibility to accommodate changes in group sizes. Smaller herds may combine some of these groups.

Special-Needs Facilities

Special-needs barns normally house sick and **transition cows.** A transition cow is a cow in the final stages of pregnancy or the first part of lactation; this transition phase is normally defined to be two to three weeks before and two to three weeks after calving and is a key period for the cow. The design of this barn should maximize cow comfort and the ability of workers to easily handle animals. Facilities to restrain animals, **drover lanes,** people-passes, proper lighting, **lifts,** tables for hoof trimming and surgery, supply storage, and calf warming boxes are often incorporated in the design of this barn.

The special-needs barn often contains freestall pens, **bedded packs,** and individual calving pens. The number and size of each pen should be based on the number of animals expected, the owner's management plan, and some additional capacity to accommodate variation in group sizes. Overcrowding of this facility is not recommended. Pens with 48-inch freestalls for pre-calving dry cows, 45–46-inch freestalls for pre-calving heifers, or bedding packs with at least 100 square feet of pack per animal are recommended. Individual pens (12 × 12 feet minimum size) or group-calving bedded packs can be used to manage cows before calving. Facilities should be designed to conveniently milk fresh cows and feed calves colostrum. Calving areas should be easily cleaned and sanitized. Gate and access lanes should be provided to easily move fresh cows and calves after calving. Sick cows should be isolated from other animals (by at least 10 feet) to prevent the spread of disease. Freestalls are acceptable for most sick cows, but bedded packs are recommended for sick cows that have trouble rising.

Milking-Cow Barn Sizing

At this time it is important to make a decision regarding which animal groups will be maintained and where each will be housed. Will milking cows be housed in a separate barn or barns? Will cows with special needs be housed with the milking herd, in a special-needs barn, or at some remote location? All these factors help determine the number of barns needed and their respective size.

Example of New Facility with Double-12 Milking Parlor

If a dairy producer plans to build a double-12 milking parlor and expects to achieve a throughput of 4.5 turns per hour, then an animal throughput of 108 cows per hour (cph) would be expected (24 milk stalls × 4.5 parlor fills = 108 cph). If the producer plans to milk three times per day and milk each group in .75 hours or less, then each pen should contain about 81 cows (108 × .75 = 81). But if the producer plans to use a parallel or herringbone parlor, then group size should be balanced so one side of the parlor can be completely filled with the last cows of a group. If the producer selects 84 cows per pen, each group could be milked in about 47 minutes (84 ÷ 108 × 60 minutes per hour = 47). Each four-pen barn would take 188 minutes (4 × 47 = 188), or 3.1 hours. Milking two barns, 8 pens of 84, would require 6.2 hours of milk time and allow 1.8 hours per shift to milk special-needs cows or to clean the parlor before starting to milk the next shift. These 8 pens of 84 milking cows ($8 \times 84 = 672$ milking cows) would lead to a herd of 830 ($672 \div .81 = 830$) animals, with dry cows, special-needs cows, and heifers waiting to calve.

Freestall Barn Design

Many different types of freestall barns can be built, and each style has its advantages, disadvantages, and costs. When evaluating barn types, consider the following characteristics: roof type, stall arrangement, and feed delivery arrangement.

Covering the freestall barn protects the animals from adverse weather conditions and allows rain and snow to be channeled away from manure areas, helping to minimize manure storage needs. Freestall barn roofs historically have been constructed with a 4:12 **roof pitch,** and over wood or steel frames, but recently hoop structures have been built and covered with canvas or greenhouse materials. Radiant heat load, initial cost, and ongoing maintenance cost should be considered when selecting between these options. Posts needed in the center of the barn to support the roof structure vary for each roof type, and their impact on feeding and freestall placement should be considered. As an example, wood barns often have posts every 12 feet of barn length, which is convenient if you want to mount freestall dividers every 48 inches but not so convenient if 46 inches stalls are desired. These support posts, if located along the **drive-by feeding** alley, provide an easy way to fasten manger stalls. Steel barns often have posts every 30 feet of barn space, which gives the barn a more open appearance. These steel structures have the additional advantage of using fewer support members, which interfere with airflow and provide roosting places for birds. Hoop structures often have no center supports, allowing complete freedom for the design of the interior of the freestall barn.

Stall Arrangement and Feeding Options

Freestall barns can be designed with rows of freestalls parallel (Figure 5-1) or perpendicular to the barn's length (Figure 5-2). Feed is normally delivered to the animals along drive-by feed mangers, which can be located inside the barn (**drive-through feeding,** Figures 5-3a–c), along the outside of the barn (drive-by feeding, Figure 5-4), or at an area away from the barn (**outside feeding,** see Figures 5-5a–b, page 69). With drive-through feeding barns, the roof protects the feed from getting wet, or dried by the sun, and the operator is protected from the elements while delivering feed, whereas this may not be true with drive-by or outside feeding. Users report greater satisfaction with drive-through barns, and cows tend to produce less in facilities with outside feeding.

Placing freestalls perpendicular to the barn's length requires less space per stall than parallel arrangements. This technique is often used if sheds are

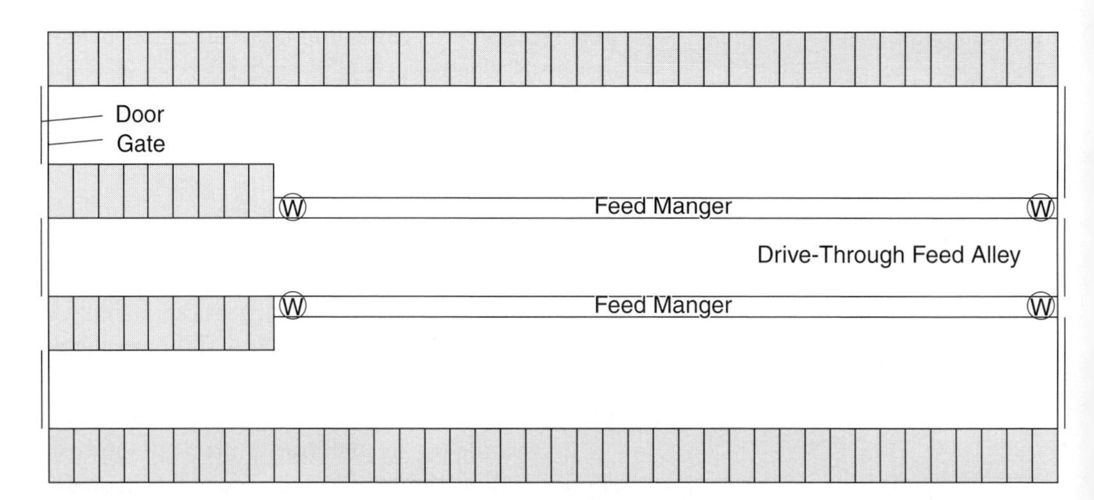

FIGURE 5-1 Two-row freestall barn with drive-through feeding (feed lane through center of barn).

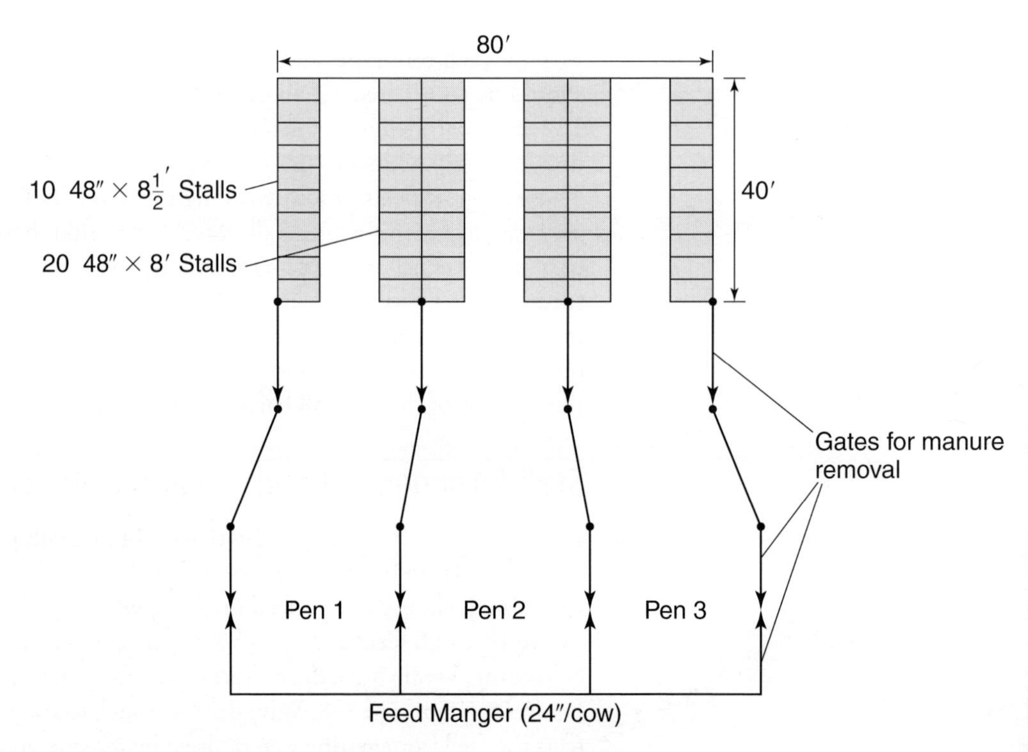

FIGURE 5-2 Three-pen freestall barn with rows of stalls perpendicular to length of barn, outside exercise lots, and outside feed mangers.

being converted and freestall numbers are to be maximized. Such an arrangement complicates manure handling, usually necessitates outside feeding, makes cow cooling more of a challenge, and complicates handling of multiple groups of milk cows; therefore, the arrangement is seldom recommended except in **retrofit** situations.

Freestall Barn Types

Most freestall barns are built with rows of freestalls parallel to the feed manger to allow manure to be easily moved to the center or end of the barn.

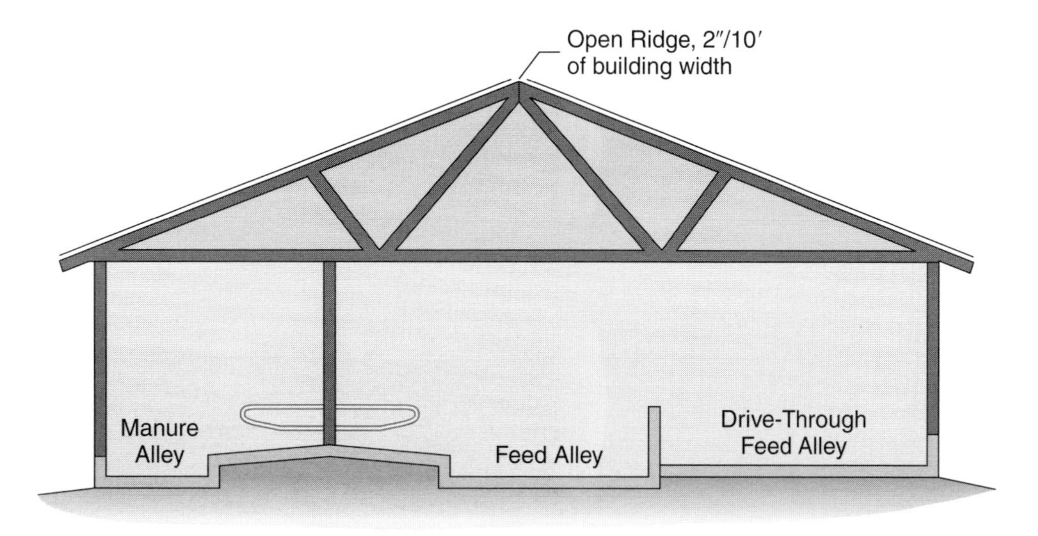

FIGURE 5-3a Two-row freestall barn with drive-through feeding; feed lane along one side of barn (cross-section).

FIGURE 5-3b Two-row freestall barn with drive-through feeding (outside view).

FIGURE 5-3c Two-row freestall barn with drive-through feeding (inside view).

FIGURE 5-4 Drive-by feeding.

Barn designs with one, two, three, four, five, or six rows of stalls are common. Increasing the number of rows of stalls in a barn increases its width and makes ventilation more difficult. Barns with one to three rows of cows placed on one side of a feed alley are narrow, easily ventilated, and inexpensive because of the rafter length, but only one side of animals uses the feed

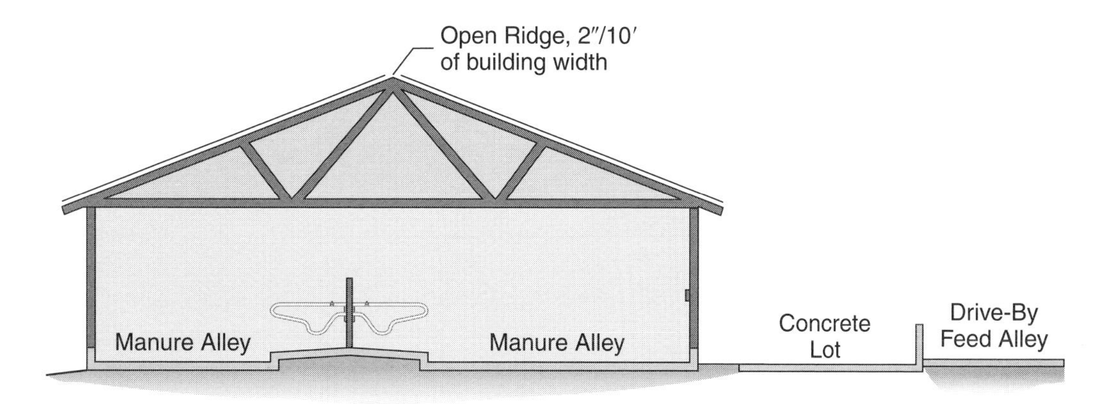

FIGURE 5-5a Two-row freestall barn with outside feeding (cross-section).

FIGURE 5-5b Two-row freestall barn with outside feeding (outside view).

delivery lane. With rows of freestalls on both sides, the feed delivery lane it is better utilized, but the barn then becomes wider, higher, more costly, and harder to ventilate. The number of rows of cows and the freestall width determine the amount of feed space provided per cow along the feed manger (i.e., placing three rows of 48-inch stalls provides about 16 inches of feed space per animal). Since barns with one row of stalls would provide more feed space than recommended, they often have feed space on only a portion of their length, and the additional length is used for additional freestalls (see Figure 5-1, page 66, for a typical floor plan).

Most new barns are two-row or three-row drive-by, or four-row or six-row drive-through. Figure 5-6 shows a three-row drive-by (the two-row is similar, but with one fewer row of stalls); Figures 5-7 and 5-8 show two different four-row drive-through configurations; and Figure 5-9 shows a six-row drive-through. The two-row drive-through configuration in Figure 5-10 is less common but has some advantages for smaller herds (again, see Figure 5-1, page 66, for a typical layout). The drive-through freestall barns normally are constructed to house four pens of cows. This aids cow flow

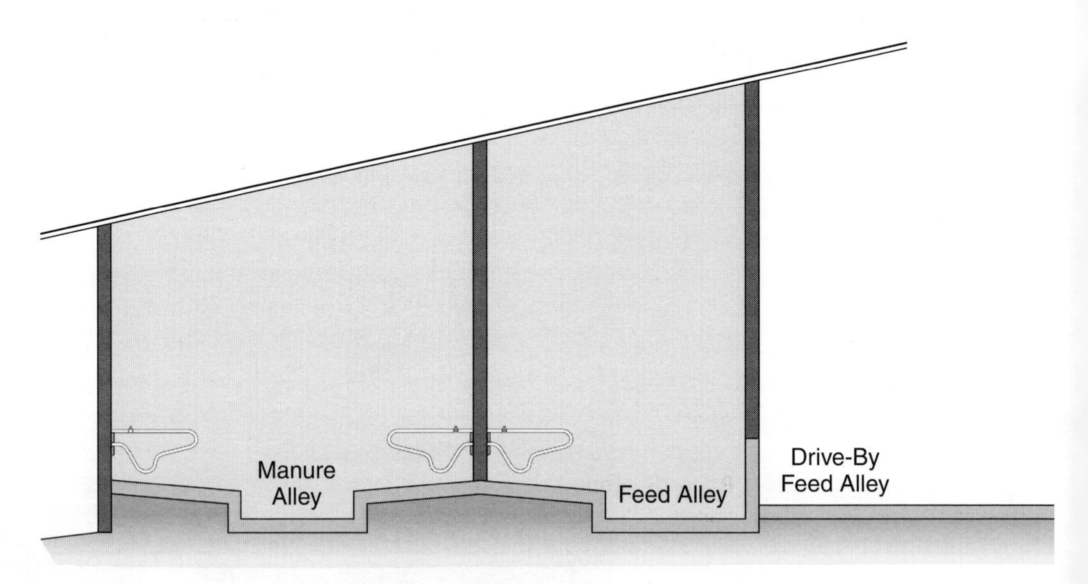

FIGURE 5-6 Three-row freestall barn with drive-by feeding and monoslope roof.

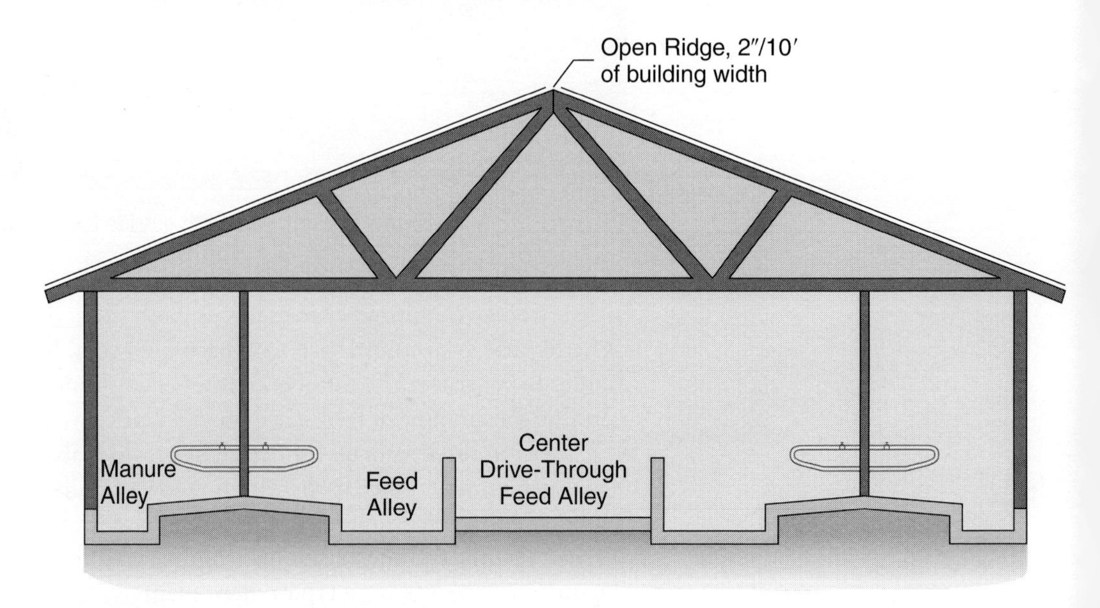

FIGURE 5-7a Four-row freestall barn with drive-through feeding and head-to-head stall design (cross-section).

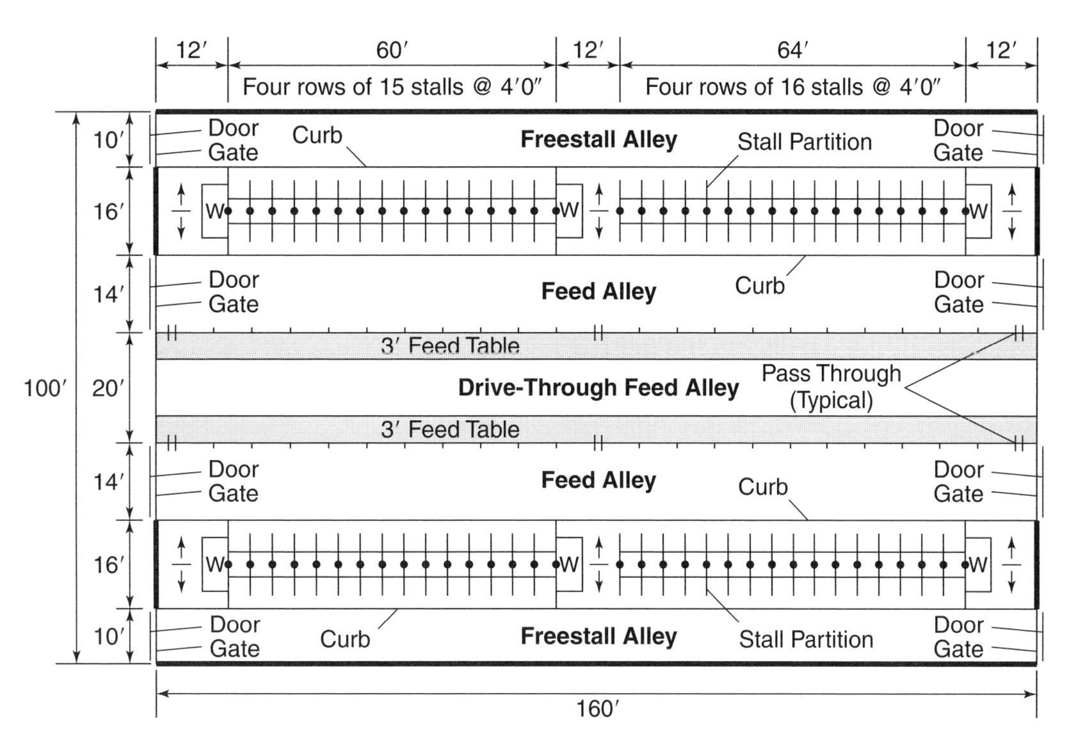

FIGURE 5-7b Four-row freestall barn with drive-through feeding and head-to-head stall design (floor plan).

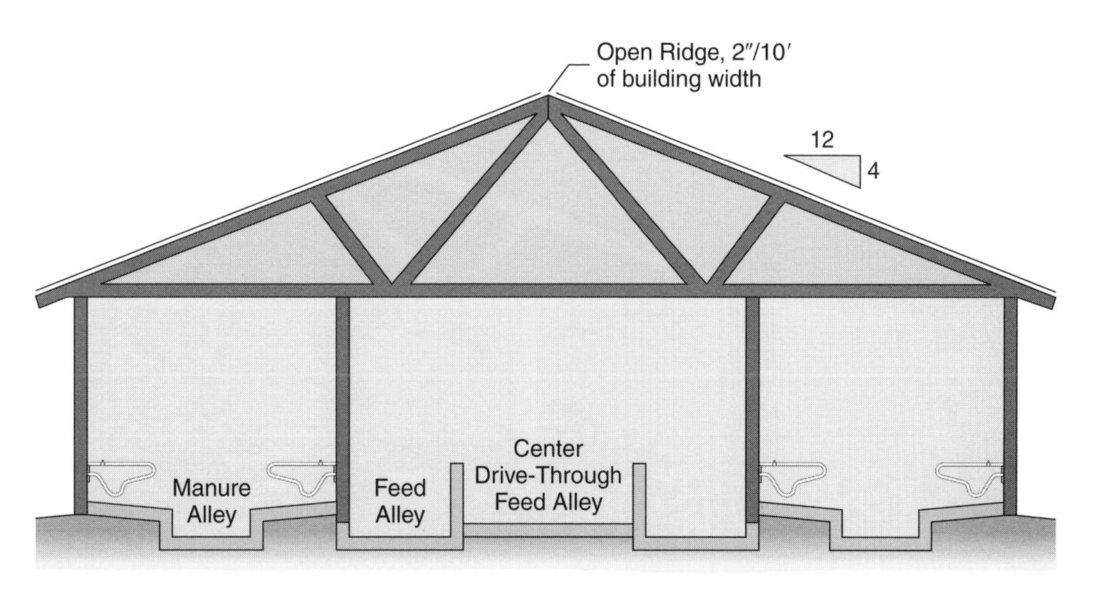

FIGURE 5-8a Four-row freestall barn with drive-through feeding and tail-to-tail stall design (cross-section).

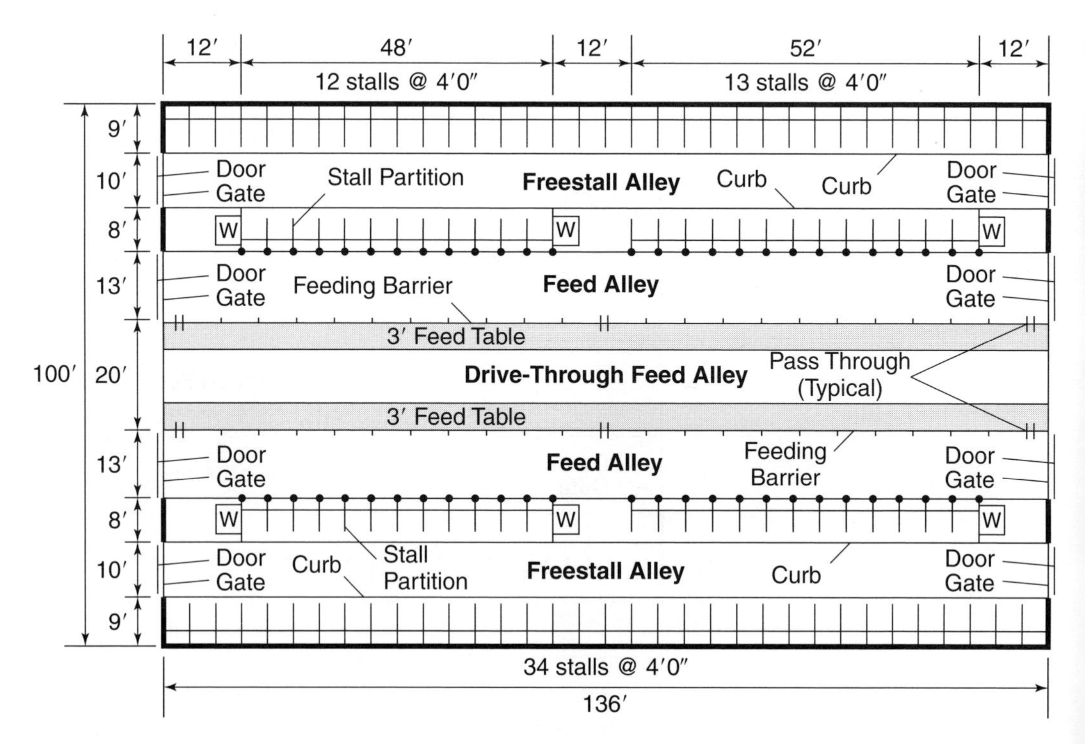

FIGURE 5-8b Four-row freestall barn with drive-through feeding and tail-to-tail stall design (floor plan).

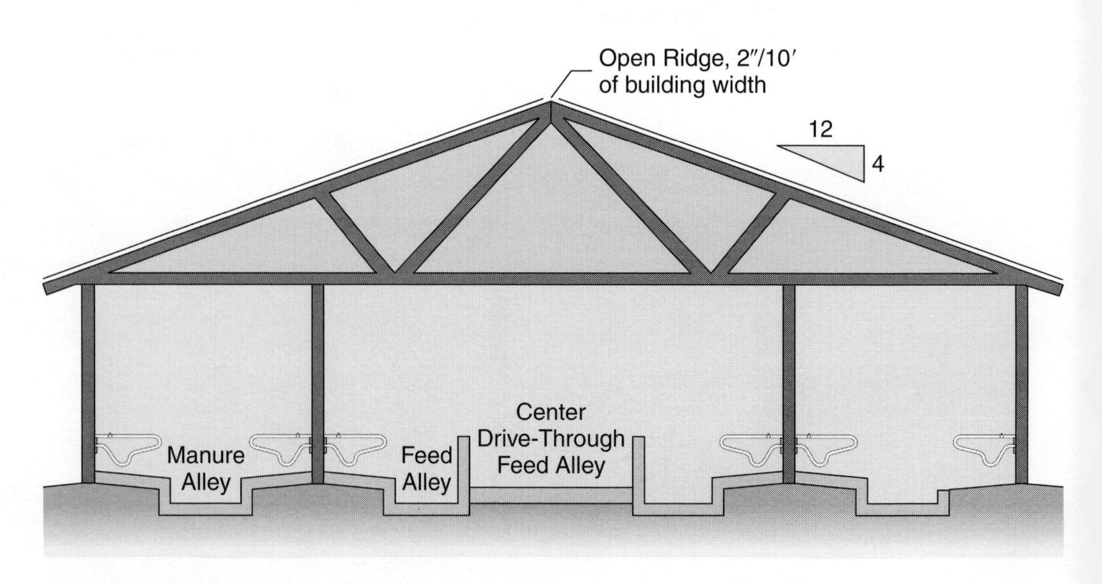

FIGURE 5-9 Six-row freestall barn with drive-through feeding.

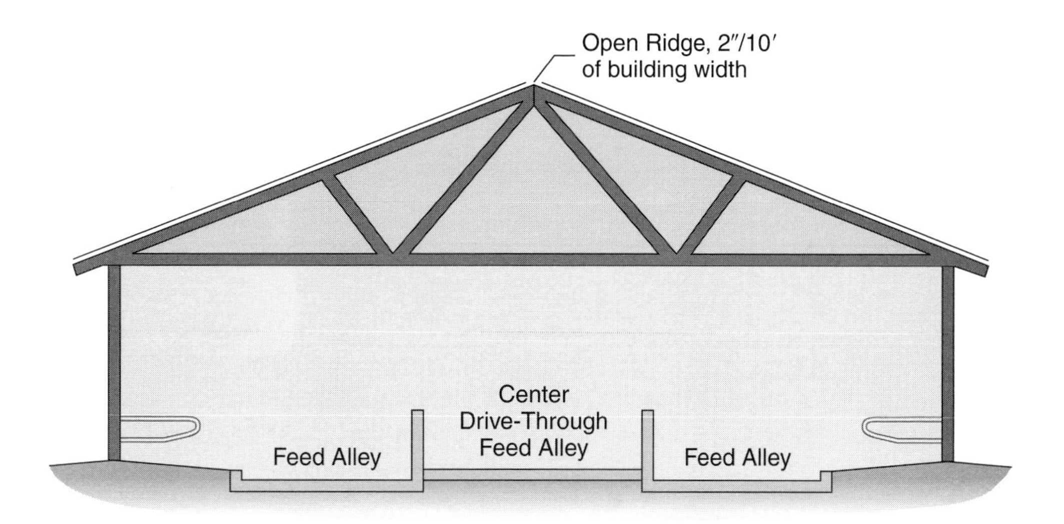

FIGURE 5-10 Two-row freestall barn with drive-through feeding (feed lane through center of barn).

because each pen of cows can exit the center of the barn to be milked and can return without interfering with any other cows. Since four pens of cows use a common lane to access the parlor, it is practical to cover this lane and protect the animals and workers from the weather. It also provides an efficient way to consolidate the manure removed from each of the four pens at one common center alley. If the two-row or three-row option is selected, only two pens of animals can access one cross alley leading to the parlor, and manure can be consolidated in a common manure pit from only two pens. In this case, covering the access lane from each barn to the milking parlor may not be cost effective. If more pens of cows are needed, additional barns must be built (which requires a larger land area than that required for a four-row or six-row barn), or pens of cows must be placed behind the primary pens, which complicates manure and cow handling. Two-row or three-row designs with drive-through feeding along one side allow better control of temperatures during cold weather, and minimize feed wastage, but increase barn costs since the portion of the barn covering the feed lane is used only by one pen of cows. (For two-row drive-through design, refer back to Figures 5-3a–c on pages 67–68; three-row design is similar, with one additional row of stalls.)

If more pens of cows are desired, or you wish to exit cows from an end of the barn, then the **back pens** of cows would need to pass through the **front pens** of cows. To make this work, cows from the front pens must be locked away from part of their housing. Often this prevents cows from lying down, increases congestion, and has the potential to increase animal stress during this time. It also requires that some form of automatic manure handling be implemented that can take manure from one pen across another pen, or multiple manure pickup points within the barn must be provided.

FIGURE 5-11 Two-row freestall barn with drive-through feeding (feed lane through center of barn).

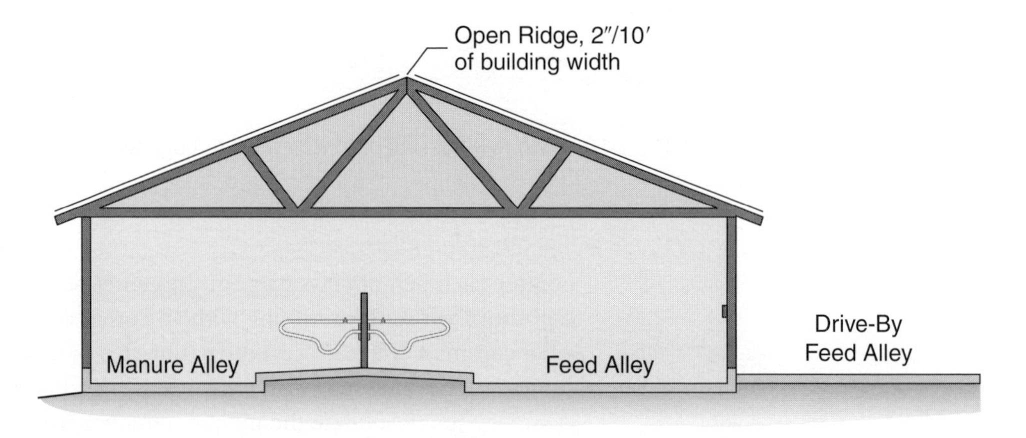

FIGURE 5-12 Two-row freestall barn with drive-by feeding and two-slope roof.

Figure 5-11 shows the inside view of a two-row drive-through barn. A row of freestalls is placed on each outside wall, and two internal rows of freestalls are located at the end of the barn. Feed space and stall numbers are balanced to ensure proper feed space per animal. The internal rows of freestalls extend into the drive alley, where feed normally is placed along a feed fence.

The freestall layouts of Figures 5-12 and 5-13 are the same, but they have different roof designs. The two-slope design (Figure 5-12) is less expensive to build than the monoslope barn (Figure 5-13), but the monoslope allows a second barn to be placed adjacent to it, making it a four-row barn. Building the second barn adjacent to the first conserves land and allows for better utilization of the feed manger lane.

Figure 5-14 shows a freestall barn with drive-by feeding in which the feed manger has been covered. Barns like these have been constructed in some cold climates to protect cows and feed from sunshine and moisture. They are not recommended, because they retard air movement, create a dungeon effort for cows, and must be cleaned manually. Because the feeder cannot observe animals as they feed, it is more difficult to identify cows needing special attention. Figure 5-15 shows a freestall barn with drive-by

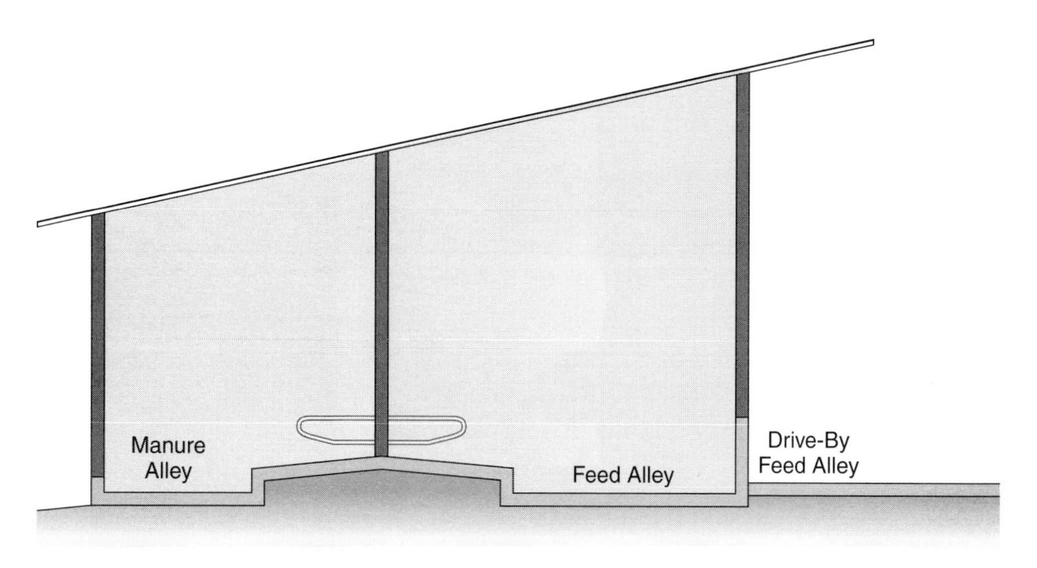

FIGURE 5-13 Two-row freestall barn with drive-by feeding and monoslope roof.

FIGURE 5-14 Two-row freestall barn with drive-by feeding and covered feed mangers.

FIGURE 5-15 Two-row freestall barn with drive-by feeding; feed space covered with rubber belts to minimize barn heat loss during cold days.

FIGURE 5-16 Three-row freestall barn with drive-by feeding; feed manger extended to increase feed space.

feeding in which the feed space has been covered with rubber belts. These belts work like freezer strips and help reduce heat loss; cows can push them aside to access feed.

Because three-row barns (Figure 5-16) provide only about 16 inches of feed space, some producers extend the feed alley as shown to achieve the desired 24 inches per cow. This extended feed alley provides additional feed space but sacrifices protection of cows and feed.

Choosing between Four-Row and Six-Row Designs

Often a manager needs to select between two alternatives when building a new facility. One classic example is the choice between a six-row barn (three rows of stalls on each side of a drive-through feed manger; Figure 5-17) and a four-row barn (two rows of stalls on each side; Figure 5-18). This is a critical decision for the producer because of the 10- to 20-year useful lifetime of the barn. (Keep in mind during the following discussion that the points are similar for the two-row versus three-row decision.)

When deciding which type of barn to build, three primary factors (the "three C's" of freestall barns) must be considered: Cost, Cow comfort, and Convenience of animal handling. In some cases secondary concerns (such as site size restrictions) also must be considered.

When estimating the cost of the freestall barn, consider the stocking rate and evaluate the price on a cost-per-cow rather than a cost-per-stall basis. Six-row barns are normally \$100–\$200 per stall cheaper to build, but higher stocking rates are recommended for four-row barns because of the feed space and animal density issues. Table 5-2 illustrates this effect: a four-row barn that costs \$200 per stall more results in only a \$67 per cow

FIGURE 5-17 Six-row freestall barn with drive-through feeding (inside view).

FIGURE 5-18 Four-row freestall barn with drive-through feeding and head-to-head stall design (inside view).

difference in initial cost if stocked 12.5 percent higher than the six-row barn. This example shows the dramatic effect of stocking rate on the decision process, in that the 20 percent higher cost of a four-row barn on a per-stall basis is only 6.5 percent higher on a per-cow basis. Another thing to think about when making this type of decision is the payback expected from the additional expenditure. Although producers must be concerned with the capital investment in their facilities, selecting the lowest initial cost may not be the best financial decision in the long term when the ongoing cost of labor is included. One way to determine whether it is worthwhile spending the additional money on a given feature is to calculate the

TABLE 5-2 Capital cost-per-cow differences for four- and six-row barns, considering recommended stocking rates.

	4-Row	6-Row	Difference
Number of cows	144	144	
Stocking rate	112.5%	100%	
Number of stalls	128	144	
Cost per stall	$1,200	$1,000	$200 (+ 20%)
Total cost	$153,000	$144,000	
Cost per cow	$1,067	$1,000	$67 (+ 6.5%)

TABLE 5-3 Calculated break-even amount of milk per day needed for each additional investment of $100 per cow.

Loan (principal and interest per year)	$15.20 per year
Repairs (5% of cost)	5.00 per year
Taxes (1.5% of cost)	1.50 per year
Insurance (.5% of cost)	.50 per year
Total cost per year per $100	$22.20 per year
Milk value	$13.50 per cwt
Milk needed ($22.20 ÷ $.135 per pound)	164 pounds of milk per lactation
Break-even milk per day (305 days per year)	.5 pounds of milk per day

expected improvement in production or labor efficiency needed to justify the expenditure. Table 5-3 shows the break-even amount of milk needed for each additional $100 spent, considering a ten-year loan repayment, additional ownership costs, and so on.

Using this example, ignoring all other factors, it would take about half a pound of milk per day to recover each additional $100 spent, or less than .4 pounds per day for the example shown in Table 5-2.

The second major area of consideration is cow comfort. Cow comfort has become the theme of the industry as it increases the expectations of dairy herds. Many herds in the United States currently have rolling herd averages in excess of 30,000 pounds of milk per year, and daily tank averages of over 100 pounds per day. These values can be achieved only by providing an environment conducive to high production. Four-row barns have two rows of cows accessing two manure alleys, whereas six-row barns have three rows of cows on the same two manure alleys. This design difference increases animal density in a six-row barn and decreases the animals' ability to get to feed and water. This congestion factor (coupled with the decreased feed space provided by the six-row option), as shown in Table 5-4, must be

TABLE 5-4 A comparison of factors relating to cow comfort for four-row and six-row barns of equal capacity.

	4-Row	6-Row	6-Row Difference
Barn length	144 ft.	105 ft.	39 ft. shorter
Barn width	92 ft.	106 ft.	14 ft. wider
Barn space per cow	115 $ft.^2$	87 $ft.^2$	24% less barn space
Barn air space per cow	1,979 $ft.^3$	1,644 $ft.^3$	17% less air space
Feed space per cow	24 in.	17.5 in.	27% less feed space

considered when evaluating the relative cow comfort factors of each design. Six-row barns are also wider, making them more difficult to ventilate; wider barns with the same sidewall height are taller and must be separated more from other buildings to ensure proper airflow, making the overall complex shorter and wider. The impact on available sites of the shorter and wider footprint of the six-row design should be considered. The third factor, convenience of animal handling, is probably the most important to address, given that any inefficiency built into the facilities will impact the business's bottom line for a long time. How you plan to sort, restrain, and treat cows for all their health and reproductive needs directly relates to labor requirements and cost. The four-row design lends itself to self-locking manger stalls because it has 24 inches of feed space per animal, whereas self-locking manger stalls are not recommended in a six-row barn because it only provides 16–18 inches of manger space, which reduces the number of animals that can eat at any one time. If self-locks are installed in six-row barns (or four-row barns with extremely high stocking levels), dry matter intake may be depressed because of the reduced number of animals that can eat at the bunk, resulting in lower production per cow. (For more information on animal management systems and their impact on dairy management, see Chapter 4, "Animal Handling Needs.") With the six-row choice, a separate area is often provided to treat and breed animals away from the freestall barn. Consider any additional costs associated with this area when comparing the cost-per-cow of each option. The task of separating animals, treating them in a different area, and then returning them to their pen often adds expense in the form of additional labor, which must also be considered. The **average annual total cost,** based on the annualized initial cost added to average operating costs, should be used to evaluate the relative cost of each option.

As mentioned, in some instances the dairy manager may want to bring cows from back pens through front pens. This is strongly discouraged with the six-row option because of the animal density that would result from three rows of cows being forced onto one feed alley. Animal density in this situation is similar to that in holding pens, which is not conducive to animal comfort.

Reported Production Differences

The 1999 Wisconsin Dairy Modernization Project surveyed producers who had expanded their herd size by at least 40 percent from 1994 to 1998. Table 5-5 shows the DHI RHA (rolling herd average) values for those farms that built new four-row or six-row freestall barns with drive-through feeding. Production was not significantly different in 1994, but was in 1998. Producers who selected the four-row option had higher production in 1994, and this production advantage increased by almost 600 pounds by 1998, resulting in more than 1,900 pounds more milk per cow per year. Table 5-6 shows that producers who built new four-row drive-through freestall barns had higher milk production, lower average somatic cell count linear scores, and higher stocking rates than producers who built new six-row drive-through freestall barns. The average price reported by these producers was very similar for the different barn types, but the six-row was slightly cheaper on a per-stall basis, and the four-row was slightly cheaper on a per-cow basis.

TABLE 5-5 DHI rolling herd average milk production values for dairies that built four-row and six-row barns when they expanded herd size.

	Number Herds	'94 RHA Milk	After Expansion	Change
4-Row	53	21,669	23,644	+1,974
6-Row	42	20,351	21,733	+1,382
Difference		+1,318	+1,911	+592

Source: Bewley, Palmer, Jackson-Smith, 2001a.

TABLE 5-6 Milk production, stocking rate, average somatic cell count linear score, and construction costs for dairies that built four-row and six-row barns when they expanded herd size.

	4-Row	6-Row	Difference
Number of herds	53	42	
1998 median herd size	245	247	
1998 rolling herd avg. milk	23,644	21,733	1,911
Stocking rate (%)	112	103	9
Average somatic cell count linear score	2.73	2.96	−.23
Cost per stall	$1,235	$1,212	$23
Cost per cow	$1,103	$1,177	−$74

Source: Bewley, Palmer, Jackson-Smith, 2001a.

Four-Row Head-to-Head versus Tail-to-Tail

When building a four-row barn, you should consider several factors in deciding between head-to-head (Figure 5-7b) and tail-to-tail (Figure 5-8b) configuration. The following is a list of characteristics of four-row head-to-head configuration barns. Much of the logic can be extended to other barn types with similar characteristics.

No cows on outside wall
- + Better air flow
- + Easier sidewall construction, curtain protection, and so on
- + Cow protected from sun and rain without need of roof overhang
- − Cold weather manure removal
- − Fewer stalls in same length barn

Cows lunge into adjacent stall
- + Need .5 feet less barn width per row of stalls (16 versus 17 feet each side)
- + Wider center crossover allows longer water tank
- − Cows' heads near each other (possible summer heat concern)

Cows can access freestalls and feed manger from common alley
- + Allows cows choice if locked onto feed manger side
- − Cows cannot be locked away from stalls
- − Manure alley on manger sides may need to be wider for stall access

Each row of stalls accessed from different manure alley
- + Two routes between feeding and resting areas
- + Cows may be less likely to interfere with each other as they exit stalls
- − Bedding of stalls may need to be done from two manure alleys

The following is a list of characteristics of four-row tail-to-tail configuration barns. The advantages and disadvantages pertain to other barn types with similar characteristics.

Cows along outside wall
- + Manure farther from outside wall (less freezing in cold climates)
- + More stalls in same length barn
- − Cows on outside wall obstruct air flow; may need taller sidewalls to compensate
- − Cost of barn roof overhang needed to protect outside row of stalls from sun and rain
- − More difficult sidewall construction, curtain protection, and so on

Cows cannot lunge into adjacent stall
- + Cows' heads not near each other (perhaps less summer heat concern)
- − Cows cannot lunge into adjacent stall; need .5 feet more barn width per row

Cows cannot access freestalls and feed manger from common alley
- + Cows can be isolated at feed manger or in freestalls
- + Cows do not need to access stalls from manger side manure alley; could be a narrower alley
- − Cows have no choice to eat or lie down if locked onto one alley

Center row of stalls narrower; smaller crossover area
- + Less space to hand scrape
- − Less space for water tank

Each row of stalls accessed from the same manure alley
- + Bedding of stalls may be done from one manure alley (possibly less labor and fewer doors)
- − More manure in stall alley than manger side (may complicate manure removal)
- − One route between feeding and resting areas (possibly more cow congestion)
- − Cows may interfere with each other as they exit stalls

Two-Row or Three-Row Drive-Through Barn Designs

Figure 5-19 shows a two-row drive-through barn design that has been used by dairy producers who want to migrate to freestall housing and plan to always maintain a small-to-medium herd size (see Figure 5-11, page 74, for an inside view). This two-row drive-through design is narrow, making it easy to ventilate, and is inexpensive to build because of rafter length. Two-row and three-row drive-through designs both have similar barn widths (60–64 feet). The advantages of the two-row configuration are that it provides two pens (one on each side of the manger) and can be designed to provide more feed space per animal. The major disadvantage of this design is that it provides only one manure scrape alley per pen, which implies that manure removal must happen when cows are out of the pen, or the configuration must be especially designed to use an alley scraper, flush manure, or slotted floor system to remove manure.

FIGURE 5-19 Two-row freestall barn with drive-through feeding.

SUMMARY

Selecting the right freestall barn design is a complicated process. The choice of freestall barn design impacts the complete design of the dairy; it defines the size of the site needed. The animal management techniques appropriate for each barn type define the amount and type of facilities and equipment in the parlor and treatment areas. For example, if a producer selects the six-row design without self-locking headlocks, then a sort-gate system that selects animals to be bred or treated and a separate area to house and treat these selected animals may be considered. If computer-operated sort gates will be used, then management must ensure that computer files are accurate and updated on a timely basis. If the sort gate malfunctions or an animal is not sorted correctly, then procedures must be in place to identify affected cows and to take corrective actions. If palpation rails or other animal batch-handling systems will be used, then the labor requirements of animal workers must be considered. Overall, the long-term **annual cost** (initial cost plus ongoing labor costs) of the facility and the cost of managing the complete dairy herd should be considered when deciding what type of freestall barn to build.

CHAPTER REVIEW

1. What aspect of a freestall most affects milk output?
2. Identify two ways in which four-row freestalls are more comfortable for cows than six-row freestalls.
3. Housing for approximately how many post-calving and early fresh cows should be provided in a 750-cow dairy?
4. Approximately how many stalls does a 750-cow dairy need for healthy milking cows, including extra stalls for transition cows?
5. Calculate how long it will take to milk each pen in a herd with 800 milking cows. Assume that the dairy producer has a double-12 milking parlor and achieves a throughput of 4.5 turns per hour. Also assume that the dairy producer has animals housed in two four-pen barns.
6. Using your answer from question 5, how many milking shifts could the dairy producer fit into one day? How long will the dairy producer have between shifts to milk special-needs cows and clean the parlor before starting to milk the next shift?
7. Examine your answers to questions 5 and 6. Are these expectations and numbers reasonable for a dairy operation? Identify potential problems the dairy operation described might experience. Identify solutions to these problems.

REFERENCES

Bewley, J., Palmer, R. W., & Jackson-Smith, D. B. (2001a). A comparison of free-stall barns used by modernized Wisconsin dairies. *Journal of Dairy Science, 84,* 528–541.

Bewley, J., Palmer, R. W., & Jackson-Smith, D. B. (2001b). An overview of Wisconsin dairy farmers who modernized their operations. *Journal of Dairy Science, 84,* 717–729.

Lester Building Systems. (1991). *Dairy free stall estimated costs.* Lester Prairie, MN: Author.

Smith, J. F., Harner, J. P., Brouk, M. J., Armstrong, D. V., Gamroth, M. J., & Meyer, M. J. (2000). *Relocating and expansion planning for dairy producers.* Manhattan: Kansas State University. (MF2424, January).

Chapter 6

Animal Housing Features

KEY TERMS

alley-scraper drive unit
cattle guard
color rendition
feedline sprinkler
feed-scraping equipment
head-lock
heat-stressed cattle
HVLS (high-volume low-speed) fan
island waterer
natural ventilation
negative pressure
quick-release latch
scissor gate
supplemental lighting
tunnel ventilation
wind shield

OBJECTIVES

After completing the study of this chapter, you should be able to

- list and describe key features of a freestall barn.
- understand how each feature affects cow comfort.
- understand how each feature affects facility operation and maintenance.
- evaluate an existing dairy operation to determine if adequate freestall features exist or can be constructed.
- incorporate construction of freestall features into a dairy modernization plan.

This chapter covers some features of freestall barns that are common across barn types and that have ramifications for cow comfort and labor efficiency. The following examples are only a subset of the total range of options available when constructing a new facility or retrofitting an existing building.

Building Access

Modern dairy housing facilities are often large because they house large herds. These large barns must be designed for easy access by human traffic and equipment for feeding, bedding, manure removal, and animal movement. In cold climates, manure may freeze during severe weather conditions and require large equipment, such as a pay loader, to remove it. Plans on how to handle this type of situation must be considered in designing the building. Entry doors and connector barns must have enough clearance to allow such equipment to enter the barn.

Access to the barn by people should minimize walking distance, save time, and minimize exposure to outside elements. Review barn plans before construction. Consider all the activities required in the dairy, and then determine if the barn supports those access requirements. For example, providing a walk lane beside the holding pen and any connector barns will enable people to pass stalled groups of cows; including normal-sized doors at the end of each barn can eliminate the need to open large overhead doors whenever entry is required (Figure 6-1).

FIGURE 6-1 This freestall barn was built with an extended door section to allow people to enter the barn without opening the large overhead door.

Ventilation

Dairy cows need a constant source of fresh, clean air to achieve their production potential. High moisture levels, manure gases, pathogens, and dust concentrations present in unventilated or poorly ventilated structures create an adverse environment for animals. Stale air also adversely affects milk production and milk quality. Proper ventilation consists of exchanging barn air with fresh outside air uniformly throughout the structures. The required air exchange rate depends on the temperature and moisture level of the outside air, and animal population and density.

Most modern dairy barns rely on **natural ventilation** to remove heat and humidity from the animals' environment. Natural ventilation of freestall dairy barns is widely used because it provides a very economical means to cool cows. A 1 mph wind is equivalent to a velocity of 88 feet per minute. Natural ventilation for a barn depends on building openings and building orientation. Barns should be oriented to maximize the airflow into the barn. In naturally ventilated freestall barns, exhausted air leaves predominantly though the sidewall during the summer and through the ridge opening during the winter. Since most barns are longer than they are wide, the length of the barn should be positioned perpendicular to the most prevalent wind during the time of the year with the highest heat patterns (i.e. if summer winds normally come from the south, the barn should be built with east-west orientation to take advantage of this wind). Openings on the sidewalls allow air to enter and escape, taking heat and humidity with it. Curtain sidewalls in cold climates enclose the barn on cold days and control the amount of air movement into the barn. Barns should never be completely enclosed; some barn openings should always be provided. Buildings should be designed with a minimum of one inch of eave opening on each side of the barn and two inches of roof opening at the top for each ten feet of width. During cold weather, curtains can be closed to this minimum amount, and then opened as temperature increases. During hot weather, the barn should be as open as possible to maximize the amount of air flowing thorough it. To accomplish this, sidewall heights of 12–14 feet are recommended. People considering higher sidewall heights should take into account the tradeoff between additional airflow and the light and heat radiated or reflected into the barn. Wide barns and barns in areas where wind directions change should be designed to be open at both ends (Figure 6-2). With naturally ventilated open-front freestall barns, an eave opening should be provided on the back wall to prevent swirling of air from the front.

Another factor to consider when attempting to maximize airflow is the proximity of the barn to other structures or land features. A high hill, wooded area, or adjacent building can serve as a **wind shield** and prevent airflow to the barn. The higher the obstructing object, the farther it should be from the barn being built. As a general rule, freestall barns are built about 100 feet from such obstructions. In many parts of the country, birds can be a problem when they enter the barn to roost and eat. Often barns are built with wooden trusses (Figure 6-3a) that provide a natural roosting place

FIGURE 6-2 Rollup doors and curtain openings on the end of a freestall barn allow for maximum ventilation.

FIGURE 6-3a Freestall barns built with an internal wood structure are often less expensive but provide places where birds can roost.

for birds, and this wood structure can also retard airflow within the barn. Other barn structure types, such as steel-framed barns (Figure 6-3b), minimize bird roosting, enhance airflow within the building, and increase visibility within the structure. Local prices, the intensity of the bird problem, and your objectives determine your choice of structure. One way to reduce bird problems is to install bird netting on the barn sides and over the roof opening, but this solution is not recommended. Birds often enter through open

FIGURE 6-3b Freestall barns with an internal steel structure provide minimal opportunity for birds to roost, and visibility is improved because of the decreased number of support posts.

FIGURE 6-4 The ridge opening of this freestall barn was covered with bird netting. During severe winter weather, such netting often becomes covered with frost and closes off the airflow needed to properly ventilate the barn and to maintain a healthy animal environment.

doors, and covering the roof opening can severely retard airflow during cold weather when frost builds up on the mesh (as shown in Figure 6-4).

Concrete Surfaces

When building a freestall barn, you must pay attention to the concrete surface that the cows are exposed to. This surface must be smooth so that the animals' hoofs are not damaged by rough edges or abrasions, and it must have grooves to prevent cows from losing their footing when they slip. After new concrete is poured, all rough surfaces should be removed before cows are exposed to it. Often a large block of concrete or a metal blade is used to remove the abrasive elements on the concrete surface. Figure 6-5 shows these rough edges being removed with a portable grinder unit. Concrete grooves are normally placed parallel to the barn's manure alleys—to allow water to flow in flush barns and to prevent scraper blades from catching the edges in barns using scraper systems. Grooves should be $\frac{1}{2}-\frac{5}{8}$-inch wide and deep, with a sharp edge to catch an animal's hoof when the animal slips. Figure 6-6

FIGURE 6-5 Abrasive portions of concrete surfaces should be removed, with a portable grinder or some other means, before animals are allowed to walk on them.

FIGURE 6-6 Concrete floors should be grooved to prevent animal slipping. Here, a new concrete floor is being grooved after the concrete has been poured and dried. This technique can ensure that each groove has sharp edges to help catch a slipping animal's hoof.

shows how this producer chose to cut groves into the concrete after it was poured and dried. This technique often results in cleaner grooves, and may not significantly increase the overall cost of the project. Care should be exercised if grooves are floated into the concrete, to ensure that the resulting grooves have a distinct edge but do not produce abrasive places that can damage an animal's hoof.

Lighting

Proper lighting is very important because it must provide the proper environment for many tasks (including office work, cleaning milking equipment, and treatment of animals) and for the health of animals. A good working environment increases efficiency, comfort, and safety. Amounts of light required will vary, depending on the tasks that are being preformed. New buildings should be designed to support the lighting needs of the dairy operation.

All fixtures should be watertight and made of corrosion-resistant materials. Wiring should be surface-mounted cable or nonmetallic conduit. When selecting light fixtures, consider initial cost, efficiency, lamp life, **color rendition,** and starting characteristics. High illumination levels (100 foot-candles) should be planned for milking parlor operator pits, offices, toilets, milk rooms, and animal treatment areas. The average light intensity in barns must be at least 15 foot-candles at cow eye level; a normal recommendation for new facilities is to install lighting designed to deliver 20 foot-candles to allow for the effect of aging and dirt buildup on light fixtures.

Research trials have shown that **supplemental lighting** can increase milk production and feed intake. The primary objective of a supplemental lighting system is to provide summer day-lengths all year. Additional light is supplied so that milk cows are exposed to a constant 16–18 hours of light and a minimum of 6 hours of darkness each day (denoted "16–18L:6D"). The expected result of supplemental lighting for commercials herd is an 8 percent increase in milk production coupled with a 6 percent increase in feed intake. Cows do not respond immediately but are expected to adapt in several weeks.

Large herds with several milking groups milked 3X often present a challenge in providing six hours of continuous darkness (Smith, Harner, and Brouk 2003). Lights may need to remain on at all times to provide lighting for cattle moving to and from the milking parlor. Low-intensity red lights may be used in large barns to allow movement of animals without disruption of the required dark period of other groups.

Dry cows benefit from a different photoperiod than lactating cows. Recent research (Dahl, 2000) showed that dry cows exposed to short days (8L:16D) produced more milk ($P < .05$) in the next lactation than those exposed to long days (16L:8D). Based on the results of these studies, dry cows should be exposed to short days and then exposed to long days post-calving.

Flexibility

A producer planning a new facility should select a barn design that supports current recommendations for grouping of animals based on stage of lactation, breeding status, production level, and so on, but the design also should be flexible enough to be used differently if the need arises. Some operators have built five-row barns to house low-producing cows on the three-row side and high-producing cows on the two-row side; others have built barns with a long end (two large pens) and a short end (two small pens) to accommodate the different-sized groups expected. Some producers have built parts of the barn with smaller stalls to accommodate heifers. All of these decisions limit the ways in which the facilities can be used in the future. One way to build flexibility into a freestall barn is to design it with equal-sized pens that can be subdivided into multiple pens. When this is done, **island waterers** are often used at the dividing points. Figure 6-7 shows an island waterer that allows cows access from each pen to a single waterer. Management can

FIGURE 6-7 Island waterers allow large pens to be subdivided into smaller pens. This arrangement allows one waterer to be shared by two pens, but increases crossover size requirements.

easily change from housing two small groups to one large group simply by removing two gates.

All of the approaches mentioned here have advantages and disadvantages. One thing to consider with the five-row option is how the roof will be designed. If each side has the same slope, rain and snow may be deposited on the feed. If the roof opening is placed at the center of the barn, each side will have a different slope. If a long-end–short-end barn is built, balancing the pens with the parlor size becomes more difficult. If pens are divided using island waterers, the back pen of animals will need to pass through the front pen to be milked.

Expandability

A building should be designed to allow future expansion. Table 6-1 shows the capacity of a freestall barn built with one end longer than the other (a long-end–short-end barn). This barn initially had two pens on the long end that could be milked with a double-8 parlor in 60 minutes, and the short end had two smaller pens that could be milked in 47 minutes. When the herd was expanded and the parlor enlarged to a double-12, only one end of the barn needed to be extended to keep the same milking time relationship with the new parlor. Planning ahead for future expansions can save money in the long term, since land preparation is most economical if done with the initial construction.

Another growth factor to consider when a new facility is planned is the manure-removal method to be used both initially and in the long term. Producers often use a tractor scraper initially, to reduce capital investment, and plan to install automated manure-removal equipment later. By knowing which automated system will be selected, you can select the proper barn

TABLE 6-1 Housing capacity (before and after expansion) of a freestall barn built with one end longer than the other.

	Initial	Expanded
Parlor size	double-8	double-12
Parlor throughput	72 cph	108 cph
One end	72 stalls	72 stalls
Milking time	60 min	40 min
Other end	56 stalls	108 stalls
Milking time	47 min	60 min
Total stalls	256	360
Herd size w/dry	305 hd	430 hd

FIGURE 6-8 Space should be provided inside freestall barns for alley-scraper drive units. This allows both the unit and maintenance workers to be protected from outside weather.

floor slope and reserve space for any additional equipment needed. In cold climates, reserve space inside the barn (Figure 6-8) to protect **alley-scraper drive units** and workers performing routine maintenance.

Gates

Gates should be designed to allow quick and easy movement of animals and people throughout the facility. Opening and closing gates can be time consuming and frustrating when properly designed gate latches are not selected. Figure 6-9 shows two different types of gate latches. The **quick-release latch** can be quickly opened from either side with one hand and the gate moved in either direction. If the gate is rigid enough, this system allows a gate to be swung toward the latch and for it to latch automatically. The chain shown would have to be manually wrapped around the post and then hooked on the bolt head provided. Any latch system (as well as people-passes), should allow a person to move easily throughout the barn with minimal time and effort.

One problem with island waterers is where to store the divider gates when not needed. **Scissor gates** that fold up and out of the way help solve this problem. Figure 6-10a shows how scissor gates are used around island waterers, and Figure 6-10b shows how scissor gates are used to form a drover lane across a freestall barn's feed alley. These scissor gates have also been effectively used for temporary or permanent separation of pens of animals.

Proper design of pens in the special-needs portion of a freestall barn allows one person to catch and restrain an animal for treatment. Figure 6-11 shows how a gate placed near the **head-lock** can be used to help direct the animal into the head-lock for treatment.

FIGURE 6-9 Gate latches, which can be quickly opened with one hand, can decrease the time spent opening and closing gates. This fiberglass gate was fitted with a quick-release latch, which is much more efficient than the chain locking device shown.

FIGURE 6-10a Scissor gates allow for out-of-the-way storage of gates when not in use. Here, multiple scissor gates are used around an island waterer to separate pens of cows or to block cows from using a crossover.

FIGURE 6-10b Here, multiple scissor gates were used along each side of a drive-through barn to form a drover lane across the feed alley. They facilitate animal movement when in the down position, then disappear after being used.

FIGURE 6-11 This bedded pack maternity area has a gate and head chute, which allows one person to corner and treat an animal.

Feed Mangers

Feed mangers should have smooth surfaces to encourage feed intake. Cows dislike an abrasive surface, and decomposed feed can accumulate on it, causing a foul smell. Smooth surfaces also allow for easy removal of uneaten portions of the diet. Constructing curbs like the one shown in Figure 6-12, which allow easy removal of feed, can improve labor efficiency and prevent buildup of feed in corners. Coating the surface (with an epoxy sealant, ceramic tile, or plastic panels) provides a smooth and easily cleaned surface. Any surface materials applied to the feed platform should be installed flush with the surface to prevent **feed-scraping equipment** from catching an edge of the material.

Most freestall barns are designed with four pens of cows that exit the barn through a center crossover alley. This requires animals to cross the

FIGURE 6-12 This feed curb was constructed with a concrete barrier that angled from the door opening to the feed alley curb. This allowed feed refusal to be easily removed from the freestall barn. The people-pass allowed easy entry into the pen; steel posts embedded in concrete helped protect the door.

FIGURE 6-13 Cattle guard used to prevent cows from entering feed alley. The boxes at the ends of the feed alley prevent feed from falling through and accumulating under the cattle guard.

drive-through feed alley. Scissor gates, electric gates, **cattle guards,** and many other systems are used to prevent cows from entering the feed alley. If cattle guards are used, they should be deep and the feed alley designed to prevent feed from accumulating under the rails. Figure 6-13 shows how the feed alley near the cattle guards can be boxed to allow cows to eat, but prevent feed from falling through the guardrails. This boxed feed area complicates feed delivery and cleaning of the feed alley, however.

Waterers

High-producing cows require large quantities of high-quality water. Each pen of cows should have at least two waterers, with a minimum of 1–2 inches of water access space per cow. Large pens should have a crossover every length of 15–20 stalls, which allows cows to move between feed alleys, waterers, and the freestall alley. A waterer should be placed at each crossover. Many different kinds of waterers are used. Commercially available models, like the one shown in Figure 6-14, are often easy to clean because of the smooth factory surface, whereas poured concrete waterers are more porous and more difficult to keep clean. As Figure 6-14 shows, water can be contaminated with manure if animals are allowed to stand in or near a waterer. Installation of a cattle guard over a tank, as shown in Figure 6-15, can

FIGURE 6-14 This commercially available water tank has a smooth factory surface and easily removed drain plug, which allow for easy cleaning. A cattle rail would prevent cows from standing in the water tank.

FIGURE 6-15 Poured concrete water tanks may cost less initially, but properly cleaning the porous surface can be time consuming and add to operating cost. The cattle rail shown helps prevent animals from standing in the tank.

prevent animals from standing in the tank, and a four-inch sanitary step around the waterer can keep cows slightly away from the tank and prevent them from defecating into it.

Fans and Sprinklers

High-producing dairy cows must be kept at comfortable temperatures. Heat stress occurs when a cow's heat load is greater than its capacity to lose heat. The heat load includes the cow's own body heat as well as external heat

FIGURE 6-16 To maximize cow comfort in warm weather, cooling fans should be placed in naturally ventilated freestall barns.

from air movement and temperature, humidity, and solar radiation. Dairy cows do not perspire heavily, so they must rely on evaporation through respiratory heat loss. **Heat-stressed cattle** have high respiratory rates that result in reduced feed intake and low rumination, which negatively affect milk production. Modern freestall barns allow cows to take advantage of shade, fans, and sprinklers to reduce heat stress. Figure 6-16 shows a new freestall barn with fans installed. Smith et al. (2003) reported that cow cooling by evaporating water from the skin surface is a very effective method of relieving heat stress and decreasing milk loss during times of high heat. The use of low-pressure sprinkler/soaker and fan systems to effectively wet and dry the cows will increase their heat loss. Dairy cows can be soaked in the holding pen, exit lanes, and on feedlines. The goal should be to maximize the number of wet-dry cycles per hour. In the summer of 2001, a study was conducted at Kansas State University to determine the effects of soak frequency and airflow on respiration rates, skin temperature, and vaginal body temperature of heat-stressed dairy cattle (Brouk, Smith, and Harner, 2003). Sixteen heat-stressed lactating cows (8 primiparous and 8 multiparous) were arranged in a replicated 8 × 8 Latin Square design. Cattle were housed in freestall dairy barns and milked 2X. During testing, cattle were moved to a tie-stall barn for a two-hour period from either 1:00–3:00 P.M. or 3:00–5:00 P.M. on eight different days in late August and early September. Average afternoon temperatures were 88°F, with a relative humidity of 57 percent. During the testing period, respiration rates were determined every five minutes by visual evaluation. Skin temperature of three sites was measured with an infrared thermometer and recorded every five minutes. Treatments were four different soaking frequencies, with and without supplemental airflow: soaking frequencies were control (no soaking), and every 5, 10, and 15 minutes; supplemental airflow was either none or 700 cfm (cubic feet per minute). Each wetting cycle provided similar amounts of water for all treatments. Initial data were collected for three initial five-minute periods prior to the start of the treatments.

Cows soaked every five minutes with supplemental airflow (5 + F) responded with the fastest and largest drop in body temperature and respiration rate, reducing the initial respiration rate by 47 percent at the end of 90 minutes of treatment. Soaking cows every five minutes without airflow (5) resulted in a similar response to soaking cows every 10 minutes with airflow (10 + F). Soaking cows every 15 minutes with airflow (15 + F) and soaking cows every 10 minutes without airflow (10) resulted in similar responses until the last 30 minutes of the study. Supplemental airflow without soaking (0 + F) resulted in little improvement over no soaking or airflow (0). Wetting had a greater effect on respiration rate and vaginal body temperature than airflow; however, the combination of wetting and airflow had the greatest effect. Respiration rates and vaginal body temperature were highly correlated. When cooling heat-stressed dairy cattle, the most effective treatment included continuous supplemental airflow and wetting every five minutes.

This data suggests that different cooling strategies could be developed for different levels of heat stress. Under severe heat stress, soaking every five minutes with fan cooling will be the most effective; under periods of moderate stress, soaking every 10 minutes with fan cooling may be adequate. Reducing soaking frequency when temperatures are lower could significantly reduce water usage. Data clearly indicate that the combination of soaking and supplemental fan cooling is superior to either single treatment. Used alone, soaking has more impact than fan cooling. The data indicate that about $\frac{1}{3}$ of the total reduction in cow respiration rates was due to airflow, and the remainder due to soaking. Under periods of severe heat stress, soaking every 15 minutes with airflow is inadequate, and soaking frequency must be increased.

Cow cooling with soaking and supplemental airflow is very effective in reducing respiration rate. Many systems may be ineffective because they do not deliver adequate water to soak the cow or have an inadequate soaking frequency. To adequately cool cows in a four-row barn, Smith recommends that fans be mounted above the cows on the feed line and above the head-to-head freestalls. If 36-inch fans are used, they should be placed no more than 30 feet apart, and if 48-inch fans are used, they should be placed no more than 40 feet apart. Fans should be operated when temperature reaches 70°F, and they should create an air velocity of 4–6 mph and airflow of 800–900 cfm per stall or head-lock. **Feedline sprinklers** should be used in addition to fans. Feedline sprinklers should wet the back of the cow and then shut off to allow the water to evaporate prior to the next cycle. Application rate per cycle should be .04 inches per feet^2, and sprinklers should operate when temperatures exceed 70°F.

High-Volume Low-Speed Fans

HVLS (high-volume low-speed) fans are configured as large diameter paddle fans with 10 blades. The blades range from 4–12 feet long, making the diameter of the fan approximately 8–24 feet. Such fans operate at speeds of 50–117 rpm (from larger to smaller diameter) and have been used in industrial buildings to circulate ventilation air at a low velocity (3 mph). They have also been used in poultry and livestock barns to provide supplemental cooling of animals by increasing air circulation and air velocity in the barn. A study conducted in several California freestall barns (Haag, 2001) used HVLS fans placed approximately 60 feet apart, mounted in the middle of the barn over the feed driveway. Research results found no difference in respiratory rates and milk production between the barns with HVLS and those with high-speed fan systems.

Kammel et al. (2003) reported that 20-to-24-foot HVLS fans installed in Wisconsin in 2001 were mounted at a height of 16–18 feet, which was typically one foot higher than the overhead garage door at the ends of the center drive-through feed lane, and were approximately 60–70 feet apart. The cost was approximately $4,000–5,000 per installed fan. Air velocities were measured at a height approximately six-inch above the cows' backs when they were lying or standing. Velocities of 200–299 fpm (feet per minute) were found over a 20-foot diameter from the center of the fans, which coincided with the feed bunk line. Air velocities of 100–199 fpm were found within 30 feet of the center of the fan, which coincided with interior freestall platforms. Horizontal velocities of approximately 100 fpm were found 40 feet from the fan center, which coincided with the outside alley and freestall platforms. Horizontal velocities in the barn were turbulent, similar to a light breeze. Air movement normally was above 100 fpm over most of the barn area (which is much less than recommended). Farmers reported improved air quality, reduced noise, drier alley floors, reduced bird populations, and less cow crowding, and they felt that the fans reduced loss of milk production during periods of high heat and humidity, compared to no fans.

Tunnel Ventilation

Tunnel ventilation is a special yet simple summer ventilation system. Its goal is to concurrently provide air velocity and air exchange in a barn. "Tunnel fans" are placed in one endwall of a building. Fans are operated to create a **negative pressure** in the barn, causing air to be drawn into the opposite endwall opening. Once in the barn, the fresh inlet air travels longitudinally through the structure and is exhausted by the tunnel fans. For tunnel ventilation to function at maximum potential, all sidewall, ceiling, and floor openings must be sealed to form the "tunnel."

Tunnel ventilation is not generally appropriate for use in cool and cold periods because it can create cold and drafty conditions. Since tunnel ventilation is a summer-only ventilation system, another means of providing air exchange must be in place the remainder of the year. Natural ventilation is the most logical choice. One concern with tunnel ventilation is that as the air that moves longitudinally through the barn, it becomes increasingly contaminated with air pollutants, and at some point the air may no longer be fresh.

Gooch (2001) states that research has shown that air movement between 400 and 600 fpm can successfully reduce heat stress in dairy cattle. The tunnel fan system for a barn should provide a total fan capacity to achieve this 400–600 fpm air velocity and 1,000 cfm exchange rate per cow. Inlets should be sized to provide a minimum of one square foot of area for every 400 cfm of fan capacity. Recommended fan controls should turn on a pre-defined band of tunnel fans when the barn air temperature reaches 65–68°F, and additional fans at 71–74°F.

Observation of tunnel-ventilated freestall barns shows that insufficient air movement may take place in the row of stalls adjacent to a completely closed sidewall. Opening the curtain wall slightly (2–4 inches) by raising the lower curtain from the bottom allows a small amount of air to enter along the length of the barn at cow level.

Since the key to making tunnel ventilation work properly is to move large volumes of air, installing a ceiling in the barn improves the performance of the tunnel ventilation system. This is contrary to the needs of a naturally ventilated barn, which uses the high ceiling area to dissipate and discharge stale air. To solve this problem, barns that will be tunnel ventilated in hot weather and rely on natural ventilation the remainder of the year can place baffles laterally across the barn at about 100-foot intervals.

Tunnel ventilation systems add measurable capital and operation costs to those of a naturally ventilated system, which must be offset by additional milk production in order for the investment to deliver a positive return. Naturally ventilated structures that provide adequate air exchange (and are outfitted with cooling fans placed over rows of stalls and the feeding area) remain the preferred system, but tunnel ventilation may be justified in new and existing barns that otherwise would provide poor cow environmental conditions.

SUMMARY

When a freestall barn is being designed, its style and features should complement the management style of the operator. Initial capital cost as well as long-term operating cost, cow comfort, and convenience of animal handling should be considered. Financial constraints will often influence which features will be selected and the best time to implement them. Each operation is different, so priorities and decisions will vary.

CHAPTER REVIEW

1. What are three features of a freestall barn that affect ventilation?
2. What is the primary objective of supplemental lighting?
3. What is the function of cattle guards and where are they often used on a dairy farm?
4. What is the most effective way to cool heat-stressed dairy cattle?
5. Identify two reasons for the ineffectiveness of cow cooling and soaking systems.
6. List the primary issues that must be considered when designing a freestall barn that will later be expanded.

REFERENCES

Brouk, M. J., Smith, J. F., & Harner, J. P., III. (2003, January 29–31). Effect of sprinkling frequency and airflow on respiration rate, body surface temperature and body temperature of heat stressed dairy cattle. In *Fifth international dairy housing proceedings* (pp. 263–268). St. Joseph, MI: American Society of Agricultural Engineers.

Dahl, G. E. (2000). Photoperiod management of dairy cows. In *Proceedings of the 2000 dairy housing and equipment systems: Managing and planning for profitability* (NRAES 129, pp. 131–136). Ithaca, NY: Natural Resources, Agriculture, and Engineering Service.

Fulwider, W., & Palmer, R. W. (2003). Factors affecting cow preference for stalls with different freestall bases in pens with different stocking rates. Abstracts from the American Dairy Science Association. *Journal of Dairy Science 86* (Suppl. 1), 158.

Gooch, C. A. (2001). *Natural or tunnel ventilation of freestall structures: What is right for your dairy facility?* (pp. 1–13). Ithaca, NY: Cornell University.

Gooch, C. A. (2003). Floor considerations for dairy cows. In *Building freestall barns and milking centers: Methods and materials* (NRAES 148, pp. 1–19). Ithaca, NY: Natural Resources, Agriculture, and Engineering Service.

Haag, E. (2001, May, 24). Cool cows for less: High-volume, low-speed fans offer huge energy savings. *Dairy Today Magazine,* 24.

Kammel, D. W., Raabe, M. E., & Kappleman, J. J. (2003, January 29–31). Design of high volume low speed fan supplemental cooling system in dairy freestalls. In *Proceedings from the fifth international dairy housing conference,* Fort Wayne, TX (pp. 243–252). St. Joseph, MI: American Society of Agricultural Engineers.

Palmer, R. W., & Bewley, J. (2000, January 29–31). *The 1999 Wisconsin dairy modernization project—Final results report.* Madison: University of Wisconsin.

Palmer, R. W. (2003, January 29–31). Cow preference for different freestall bases in pens with different stocking rates. In *Proceedings from the fifth international dairy housing conference,* Fort Wayne, TX (pp. 155–164). St. Joseph, MI: American Society of Agricultural Engineers.

Smith, J. F., Harner, J. P., & Brouk, M. J. (2003, July 9–10). Dairy facilities—Putting the pieces together. In *Four-state applied nutrition and management conference,* Lacrosse, WI (pp. 34–45). MWPS-4SD16. Ames, IA: Midwest Plan Service.

Smith, J. F., Harner, J. P., Brouk, M. J., Armstrong, D. V., Gamroth, M. J., & Meyer, M. J. (2000, January). *Relocating and expansion planning for dairy producers.* Manhattan: Kansas State University.

Chapter 7

Freestall Design and Bedding Materials

KEY TERMS

brisket board
culling rate
freestall divider
head-to-head barn
head-to-tail barn
mattress-based stall
neck rail
rubber mat
sand-based stall
tail-to-tail barn
waterbed

Objectives

After completing the study of this chapter, you should be able to

- understand how freestall design affects overall facility maintenance.
- understand how freestall design affects the positioning and comfort of a cow.
- understand the advantages and disadvantages of various bedding types.
- decide which freestall design and bedding material will work best in an existing dairy, and incorporate decisions into the operation's modernization plan.

This chapter discusses the actual design of freestalls and bedding materials. The main objective in building freestall barns is to provide a labor-efficient, clean, comfortable environment. Freestall dimensions, divider design, base material type, and bedding material choices are critical in accomplishing this objective. The type of animals being housed should determine free-stall specifications, and the bedding material chosen will influence manure handling.

Freestall Design

A cow will use a correctly sized freestall because it is easy for the animal to get up and down in, and it provides a comfortable surface to lie on. To support labor efficiency, stall size should encourage animals to lie straight in the stall, with their rump over the back of the stall (so that manure will fall in the manure alley and not on the stall surface; Figure 7-1). The size of a freestall is determined by the animal's size. Several publications, such as the Midwest Planning Service *Dairy Freestall Housing and Equipment* handbook (*MWPS-7*), provide the correct dimensions for each animal size; selecting a size that accommodates the larger animals in a group is recommended. Figure 7-1 also shows the interior of a **head-to-tail barn,** where all rows of cows face the outside walls of the barn. The owner of this barn reported that he preferred this arrangement because he could observe the rear of all cows lying as he walked down the feed alley.

The key freestall dimensions to consider are curb height, stall width, stall length, **neck-rail** height, and **freestall divider** mounting specifications. If curbs are too low, manure may enter the stall when being removed from the barn; if too high, cows will be reluctant to back out of the stalls. A curb height of 10 inches is recommended (normally it will be 9.5 inches if

FIGURE 7-1 Freestalls should be designed so that cows lie straight in the stall with their rump extending over the curb.

a 2- × 10-inch plank is used to form the curb. Stalls should be wide enough to allow animals to recline and rise easily. If stalls are too wide, animals will tend to lie at an angle, and may even lie backwards in the stall. Both of these situations can lead to dirty cows, and require additional labor to clean stalls because animals will deposit manure on the stall surface. For the average mature Holstein herd, 45–46-inch-wide stalls often meet these requirements the best. Larger stalls, 48 inches wide, may be considered for extremely large milking cows or pregnant dry cows. Often, 48-inch stalls are built as a convenience to the builder, whereas 45–46-inch stalls would offer the advantages mentioned, plus allow 6 percent more stalls per barn.

Cows prefer to lunge forward when rising, because transferring their weight forward allows them to lift their hindquarters more easily. Eight feet of effective stall length is recommended for mature Holstein cows. Actual stall length can be as little as seven feet if the stall design allows the cow to lunge into an adjacent stall. Short stalls are not recommended but can be used when retrofitting existing barns with limited space. Lunging forward into an adjacent stall is recommended to encourage animals to lie and rise straight in their stalls. Often a **brisket board** will be mounted in the front of the freestall to help position the animals when lying down and to provide a bracing point for cows when they rise. Brisket boards are recommended to be installed 66–71 inches from the manure curb for mature Holstein cows.

Freestalls should be designed so that cows can easily rise. Proper positioning of the neck rail is critical in stall design. The neck rail functions as a guide to position the cow correctly when she enters the stall or stands in it before or after rising. If the neck rails are too low, cows may be reluctant to enter the stalls, may stand half in and half out of the stalls, and may have difficulty getting up. To identify if stalls are correctly designed, watch cows as they attempt to rise. Figure 7-2 shows a cow having trouble getting up. This posture is often seen in barns in which the freestall neck rail is mounted too low. If the rail is too high, animals may not position themselves correctly in the stall before or after lying down. The normal recommendation is for the neck rail to be mounted at least 45 inches above the surface on which the cow stands. This dimension must take into account the height of any bedding or bedding mattresses used.

Anderson (2002) discusses the characteristics of freestalls that contribute to cow comfort. Based on work in Europe, Canada, and the United States, he reports that longer stalls, loops with wider openings, higher neck-rail placement, brisket boards no more than four inches above the stall bed, and placement of the platform in front of the brisket board at approximately the same level as the bed, are important factors in improving cow comfort and stall usage. Proper freestall divider mounting depends on the type of divider selected. When making a buying decision on a freestall divider, carefully consider its length, the space between top and bottom rails, and its shape. Visiting farms, observing the combination of stall size and loop design, and noticing how animals actually use their freestalls should make you comfortable with your choice. Be cautious about allowing a

FIGURE 7-2 Watching cows lay down and rise in freestalls can help in determining whether the stalls are designed correctly. If cows place their heads over the neck rail when they rise, it usually indicates that the rail is mounted too low.

builder to make these choices. If someone recommends mounting freestall dividers differently from the manufacturer's recommendations, then you may be selecting the wrong divider type.

Freestall Divider Design

There are many freestall divider designs currently on the market, often referred to by descriptive names such as side-lunge, wide loop, straight loop, and so on. Whichever stall divider type is selected, its length should allow a 14-inch space between the end of the divider and the manure curb when mounted (Bickert, Holmes, Janni, Kammel, Stowell, Zulovich, 2000). Additional space may encourage a cow to enter another cow's space, and less space may result in cows hurting themselves if they hit the divider as they enter the stall. Remember that if barns have different stall lengths, appropriate stall dividers should be selected for each.

A critical dimension to consider when selecting a divider is the distance from the top of the stall surface to the bottom of the neck rail—which should be 45–50 inches, with an absolute minimum of 42 inches. Fulwider and Palmer (2003) showed that the percentage of time that cows lie in a stall increased significantly when the neck rail was raised from 45 to 50 inches in a barn with **mattress-based stalls** (Table 7-1). This research was conducted in a new freestall barn in which half of the stall dividers were replaced with a different loop design that allowed the neck-rail height to be

TABLE 7-1 Effect of changing neck-rail height on the percentage of stalls with cows lying in them.

	Before (1-29-03 to 2-26-03)	After (4-03-03 to 5-01-03)
Average stocking density	96%	94%
Average temperature (THI)	21.8°F (26.9)	49.1°F (49.9)
Neck rail (45" before and after)	42.1%[b]	43.8%[b]
Neck rail (45" before, 50" after)	40.0%[b]	51.4%[a]

[a,b]Percentages with different superscripts differ ($P < .05$).

increased and the mounting bar at the front of the stall to be removed. The removal of the mounting bar at the front of these 8-foot **tail-to-tail barn** stalls may have been an important reason for the increased stall usage; if a cow hits an obstruction when attempting to rise, it will discourage the use of the freestall. The stall divider type may also have had some effect on stall usage if it provided a more comfortable stall environment.

Another important dimension is the distance from the top of the stall-divider bottom rail to the stall-base surface. If the stall divider provides sufficient space for the animal's head, then the bottom rail must be high enough to discourage the animal from crawling over it. Producers have reported dissatisfaction with extremely wide loop designs because cows get jammed in them, and they tend to encourage cows to lie at an angle in the stall. Field observation suggests that this bottom rail should be at least 10–12 inches above the stall surface. Another consideration is the amount of space provided at the rear of the stall underneath the divider. Extra space here encourages cows to lie with their rump under the divider, and results in cows lying at an angle in the stall. Dividers that are mounted perpendicular to the stall-mounting surface and protrude at least 12 inches past the brisket board are recommended. Research is being conducted to improve stall divider design, which makes it important for users to consult current research results before making a buying decision.

Barns with rows of **head-to-head barn** stalls allow animals to lunge into the stall in front of them. This feature saves space but can also lead to animals being caught between neck rails if they try to escape through the front of the stall. If this happens, a cable, strap, or pipe may be required between the rows of stalls. Such a device must be high enough to allow the cow to lunge forward but low enough to prevent her from entering the adjacent stall.

Troubleshooting Freestalls

Neck rails should be designed to be easily adjusted. Initially the neck rail should be mounted over the base of the brisket board or 66–71 inches from the manure curb. After the animals have had time to adjust to stalls, the neck rail may need to be adjusted backward or forward, depending on the behavior observed. Cows prefer to place all four feet on the stall surface before reclining. If a high percentage of the animals stand half in and half out of the stalls before lying down, the neck rail may need to be moved forward. If the animals are standing in the stall, but so far forward that they defecate or urinate on the stall base, then the neck rail may need to be moved back. Care should be exercised when making such adjustments, because a few inches can have dramatic effects.

Bedding Material Choices

Freestalls are often thought of as having two components, a stall base constructed of clay, concrete, wood or some other material, and a bedding surface. With deep **sand-based** freestalls, sand fulfills both functions. Whatever components are selected, freestalls should conform to the shape of the cow when she is resting and provide cushion when she is reclining and traction when she is rising. It is recommended that the stall surface be 3–4 percent higher in the front than in the back of the stall. This discourages forward movement while resting and improves stall drainage; also, many people feel that cows actually prefer to lie uphill. Avoid stalls that are lower in the front than the back; cows have difficulty getting up in these circumstances. Excessive stall slope may also cause cows to lie incorrectly in their stalls, as shown in Figure 7-3.

FIGURE 7-3 Even new freestall barns can be uncomfortable if not designed correctly. Too much slope in the stall may cause cows to lie at an angle.

Many different base-and-bedding combinations have been tried over the years, with varying costs and results. Cows dislike concrete-based stalls unless a thick bedding surface is maintained on top of them. Straw, sawdust, manure solids, and other organic bedding surface materials have been used successfully over concrete bases, but their cost is sometimes prohibitive. Wood-based stalls have not been successful, because the wood rots and gets slippery when wet. Clay-based stalls can provide cow comfort but require a large maintenance effort because cows dig large holes in the front of the stalls. Producers have used rubber tires for freestall bases; cows seem to like these tire-based stalls, and bedding requirements are decreased, but getting the tires installed properly is very important. Tires should be of the same size, placed tightly together, and carefully packed with material to hold them in place. Different types of **rubber mats** have been tried over the years with mixed results. Some get slippery and promote hock damage, and others have deteriorated in a short period of time.

Mattress-based stalls currently are very popular, and for most producers the choice of freestall bases is between sand, mattresses, or **waterbeds.** Mattress-based stalls normally have rubber particles or other filler that conforms to the animal's body and may offer an insulating effect during cold weather. They have a cover that provides animal traction, may be waterproof, and is durable enough to withstand animal traffic. The initial cost of mattress-based stalls is normally $50–100 per stall, and their expected useful life is between four and seven years. Mattress-based stalls must have some type of absorbent bedding applied to them, but the amount is less than deep-bedded stalls over concrete. The initial investment in sand-based stalls is low, but the labor to fill and maintain them, the cost of the sand used, and the adverse effects of the sand on manure handling and storage result in a high maintenance cost. Waterbeds consist of two layers of vulcanized rubber filled with water and calcium chlorinate (to prevent freezing during cold weather). Research has shown that stall use initially is lower with waterbeds than mattresses. The movement of the water inside tends to scare cows, but use increases over time as cows become accustomed to them. Newer models have decreased this movement by placing baffles inside the rubber bladder, which appears to increase cow acceptance.

Sand versus Mattresses—Performance and Producer Satisfaction

A recent survey of Wisconsin producers who increased herd size by at least 40 percent from 1994 to 1998 showed no significant difference in DHI milk production or somatic cell counts between those using sand and those using mattresses after their expansion (Palmer and Bewley, 2000; Table 7-2). Producers using sand seemed to be more satisfied with cow comfort and less satisfied with manure management and bedding than those using

TABLE 7-2 Average production and satisfaction values for herds using mattresses and sand bedding (Palmer and Bewley, 2000).

	Mattresses	Sand
Number of herds	69	145
DHI 1998 RHA milk (lb)	22,519	22,539
Average linear SCC	2.88	2.80
Culling rate (%)	34	32
Cow cleanliness*	4.12	4.47
Hock damage*	4.22	4.72
Bedding use and cost*	4.25	3.95
Manure management*	4.32	3.43

*Average satisfaction reported on a scale of 1 (very dissatisfied) to 5 (very satisfied).

mattresses. Sand users reported significantly higher satisfaction scores for cow cleanliness and hock damage, whereas mattress users reported significantly higher satisfaction with bedding use and cost, and manure management. **Culling rates,** although not significantly different, showed a slight numeric advantage among sand users. Results of this survey can be found on the University of Wisconsin Dairy Science Department's Web site (www.wisc.edu/dysci).

An Iowa study, which was designed to evaluate six different freestall surfaces, found that stalls ranked differently by week of trial, with cow preference switching between sand and mattresses (Thoreson, Lay, and Timms, 2000). Sand ranked highest in the summer, but usage declined from summer to winter.

Other research conducted in Europe demonstrated that cows showed definite preferences for some types of mattresses and that cow preferences changed over time (Sonck and Daelemans, 1999). It was suggested that cows need time to adjust to some types of mattresses, and other mattresses get harder and less comfortable over time.

Table 7-3 gives the results of a study (Palmer, 2003) that showed that stall base type affects cow preference. This study (Experiment 1) reported the stall usage for a four-row freestall barn with a 100 percent stocking rate. Observations of cows lying or occupying stalls (standing or lying) were recorded for a nine-month period. Sand and mattress-I (rubber-filled) stalls consistently had higher stall use percentages; concrete and soft rubber

TABLE 7-3 Experiment 1: Cow preference for different stall-base types (for a four-row barn with 100 percent stocking).

	Soft Rubber Mat Type I	Waterbed	Mattress-I (rubber-filled)	Mattress-II (foam-filled)	Concrete	Sand	Average
Lying	32.9%	45.4%	65.2%	57.4%	22.8%	68.7%	51.0%
Standing	24.6%	7.9%	17.0%	20.7%	8.8%	3.3%	12.1%
Occupied	64.8%	61.6%	88.3%	84.1%	38.7%	79.0%	70.1%
Number of observations	6,727	6,727	6,727	6,727	7,688	13,454	

mats consistently had the lowest percentages; mattress-II (foam-filled) and waterbed percentages were intermediate. The sand-based stalls had the highest overall lying percentage, but mattress-I and mattress-II had the highest occupied stall percentages. Cows appear to prefer to stand on soft surfaces provided by mattresses or soft rubber mats rather than sand stalls or concrete alleys. The lying percentage advantage of sand over mattress-I (3.5 percent) was small compared to the occupied stall advantage of mattress-I over sand (9.3 percent). This suggests that cows like to lie down on both stall bases, but prefer to spend nonlying time standing in mattress-I stalls rather than on concrete manure alleys. Some stall base types were consistently inferior to others. Lying percentages for concrete and soft rubber mats were always below the average lying percentages. Mattress-I stalls consistently ranked higher than mattress-II for lying and occupied-stall percentages, which indicates that not all mattresses are equally desirable to cows (therefore, general statements about mattresses may be misleading). The length of time that cows are exposed to the different stall bases affects lying and occupied percentages. The waterbed-based stalls required a longer adaptation time, whereas use of soft rubber mat stalls in this trial decreased over time.

Table 7-4 shows the results of Experiment 2, conducted in the same barn as Experiment 1 (Fullwider and Palmer, 2003). Two different mattress types and three different soft rubber mat types replaced the sand, concrete, and waterbed stall bases. Cow preference was strongest for foam- and rubber-filled mattresses. Cow preference for the two mattress types previously tested—which had been preferred—now was intermediate. These two mattress types were installed approximately three years before the other stall bases, so it is impossible to determine whether the new mattress types were superior or the decrease in cow preference for the existing two types was due

TABLE 7-4 Experiment 2: Cow preference for different stall-base types (for a four-row barn with 100 percent stocking).

Stall Base	Lying (%) 6-19/12-17	Lying (%) Experiment 1	Occupied (%) 6-19/12-17	Occupied (%) Experiment 1
Mattress-III (foam-filled)	62^{a}		91^{a}	
Mattress-IV (rubber-filled)	59^{ab}		84^{b}	
Mattress-I* (rubber-filled)	57^{b}	65	85^{b}	88
Mattress-II (foam-filled)	52^{c}	57	81^{b}	84
Soft rubber mat type II	51^{c}		73^{c}	
Soft rubber mat type III	43^{d}		64^{d}	
Soft rubber mat type IV	42^{d}		65^{d}	
Average	52	50	78	75

^{a,b,c,d}Percentages within rows, lying and occupied analyzed separately; different superscripts differ ($P < .05$).

to an aging effect. Rubber mats were consistently the least used. Differences in stall usage existed between different manufacturers' foam- and rubber-filled mattresses. Visual inspection showed differences in deterioration and consistency of surface level of the different products over time. These factors can influence the life expectancy of each product and should be considered along with cow preference when making a buying decision.

SUMMARY

Proper freestall design is a critical part of the planning process for any dairy producer building or updating a freestall barn. With the high production levels expected of a modern dairy cow, every effort must be made to enhance her comfort. Critically looking at existing facilities and observing freestall usage on different farms should help the producer select the correct stall dimensions and freestall divider type.

The choice of bedding base and bedding materials should be based partly on the availability and costs associated with each system. There is no ideal combination at this time; each system has both advantages and disadvantages. Most producers in the Midwest will choose between sand-based and mattress-based freestalls; both work when managed properly. There are many different types of mattresses available at any given time, so each producer must investigate what is currently available. Visit farms that have installed the type of bedding that you are considering, and observe its wearability and cow use. The decision of which system to use should be made early in the planning process because it has such a far-reaching impact on manure handling and other facility-related decisions.

CHAPTER REVIEW

1. Identify the five significant design dimensions of a freestall.
2. What are three indications that a freestall was designed and installed correctly?
3. What are the three primary indications that a freestall neck rail is too low?
4. What is the purpose of a freestall divider?
5. How does an improperly installed freestall affect facility maintenance?
6. Identify the three reasons for which bedding is important in a freestall.
7. List the advantages and disadvantages of both sand bedding and mattress bedding.

REFERENCES

Anderson, N. (2002, November 12–13). *Cozying up to cow comfort.* Handout, Midwest Dairy Herd Health Conference, Middleton, WI. Madison: University of Wisconsin—Extension.

Bickert, W. G., Holmes, B., Janni, K., Kammel, D., Stowell, R., & Zulovich, J. (2000). *Dairy freestall housing and equipment handbook* (7th ed.; MWPS-7). Ames, IA: Midwest Plan Service.

Fulwider, W., & Palmer, R. W. (2003). Factors affecting cow preference for stalls with different freestall bases in pens with different stocking rates. Abstracts from the American Dairy Science Association, *Journal of Dairy Science 86* (Suppl. 1), 158.

Palmer, R. W. (2003, January 29–31). Cow preference for different freestall bases in pens with different stocking rates. In *Proceeding from the fifth international dairy housing conference,* Fort Wayne, TX (pp. 155–164). St. Joseph, MI: American Society of Agricultural Engineers.

Palmer, R. W., & Bewley, J. (2000). *The 1999 Wisconsin dairy modernization project—Final results report.* Madison: University of Wisconsin.

Sonck, B., & Daelemans, J. (1999, August 22–26). Comparison of free stall cattle mattresses in a preference test. In *Proceedings of the 50th Annual Meeting of the European Association for Animal Production,* Zurich, Switzerland. Wageningen, Netherlands: Wageningen Pers.

Thoreson, D. R., Lay, D. C., & Timms, L. L. (2000). *Dairy free stall preference field study.* Ames: Iowa State University.

Chapter 8

Site Selection

KEY TERMS

barn orientation
geological homework
gravity-flow manure handling system
nonpoint pollution source
nuisance problem
nutrient budget
open-sided barn
point pollution source
prevailing wind
shrink–swell potential
silage leachate
site elevation
site selection
sun angle
three-phase power
topographic survey

OBJECTIVES

After completing the study of this chapter, you should be able to

- list and describe key issues affecting dairy site selection.
- identify the advantages and disadvantages of retrofitting an existing dairy.
- identify the advantages and disadvantages of building a new dairy.
- evaluate existing or potential dairy sites using the site-selection criteria.

Site selection is one of the most important aspects of the dairy planning process. A good site will increase cow comfort and labor efficiency, and support the long-term growth of the operation. Each site has its own size and orientation requirements and limitations. You should research the different options available, their limitations, and their implications for the long-term growth of your business.

The principles of site selection are similar for producers planning a new dairy and those transitioning from an existing dairy. Producers transitioning from an existing dairy must decide whether to build a new facility at a different site or to add facilities to an existing site. This choice is critical and should be made before any additional resources are deployed at the old site.

Long-Term Goals

Proper site selection should conform to the current *and* long-term goals of the operator, and consider land required for facilities, crop production, and manure disposal. No matter what size of operation a producer is considering, facility site selection should support potential future growth. For example, if a producer currently milks 60 cows in a tie-stall barn with a pipeline milker and wishes to milk 120 cows housed in freestalls with a parlor capable of milking 60 cows per hour, the selected site and building arrangement should support the number of cows that the parlor can accommodate. In this case, site selection and building placement should not consider just the 120-cow herd size, but the potential herd of 390 if milking 3X (60 cph × 6.5 hour shift) or 630 if 2X (60 cph × 10.5 hour shift). This thought process is difficult for many producers considering their first expansion, but experience has shown that once a dairyman switches to freestall–parlor type systems, successive expansions come quickly. Even if the producer has no long-term plans to expand further, the next owner may, and it would be foolish to exclude the possibility.

Herd-Size Implications of Expansion Alternatives

It is often difficult for producers to choose the correct herd size for the initial and subsequent phases of an expansion. Starting with a smaller initial herd size can be an advantage in that (1) fewer additional employees must be hired, allowing the owner more time to adjust to an expanded management role; (2) less initial investment is required; and (3) less land and equipment are needed for feed procurement and manure management. The most difficult problem associated with small expansions is the cost to milk each cow. Normally the cost of housing and of the milk cow herself do not vary by the size of operation, but since parlor costs come in fixed increments, the associated cost per cow to harvest milk can vary substantially based on the amount of time the parlor is used. A recent summary of the cost of new milking parlors in Wisconsin (Frank, 2002) shows an average cost of about $18,000 per milk stall. (This value includes the cost of new equipment, and the building to

TABLE 8-1 Equipment and labor cost to harvest milk for parlors at differing levels of use.

	Full Use	Half Use	Quarter Use
Cows milked per milk stall	30	15	7.5
Parlor investment per stall per cow	$600	$1,200	$2,400
Repayment cost ($/cow/yr)*	$116	$232	$463
Parlor labor cost ($/cow/yr)*	$218	$233	$264
Total cost ($/cow/yr)	$334	$465	$727
Cost per cwt to harvest milk	$1.67	$2.33	$3.63

*Based on 9% interest and 7-year repayment, 20,000 lb annual milk production, and $10/hr labor.

house the parlor, equipment, offices, and holding area.) Producers tend to build double-8 or larger parlors that offer better use of milking labor than smaller parlors, and must decide if the herd size being considered can support the cost of such a new parlor (double-8, 16 stalls at $18,000 per stall, or about $288,000). Since most producers who build new milking parlors milk three times per day, each milk stall has the capacity to milk about 30 cows (6.5 hour shift × 4.6 turns per hour). Table 8-1 shows that the cost to harvest milk decreases by almost $2 per cwt ($1.67 versus $3.63) when a parlor is used at its full potential. Producers considering moving to a parlor system should remember that their competition, who fully utilize their parlors, have about $600 investment per cow ($18,000 per 30). To put this in perspective, if a producer wants to milk 120 cows and have an equivalent investment per cow, then a low-cost parlor option totaling about $72,000 should be considered (120 cows × $600).

Site Selection Factors

Location

Proper location selection can maximize cow comfort, save on construction costs, and reserve space for future growth. Do not select a site because the land in question has no other good uses. A team approach to site selection, which allows many ideas and points of view to be incorporated, often results in the best final decision. Remember to consider safety and security when selecting a site; calves housed in unlit areas near infrequently used roads may disappear at night.

Climate

Local weather patterns (temperature and humidity ranges, rainfall amounts and timings, and winter storm frequency and severity) will govern the type of facility that should be constructed on a site.

Site Size

To determine the area required, the operator should sketch an overview of the proposed dairy, showing the location of all barns, the milking facility, feed storage, and manure storage. Existing buildings, roads, streams, property lines, utility lines, drainage ways, wells, neighbors, and any other sensitive areas should also be shown. Barn dimensions can be calculated based on the type of freestall barn and the number of stalls desired. Adequate space should be provided so that no structure interferes with the airflow of other structures. Large freestall barns normally must be separated by about 100 feet.

Manure storage should be calculated considering expected herd size, milk production level, and any additional manure system requirements. Water from flush systems and sand from sand-based freestalls can increase the storage requirements and, obviously, the manure pit size. Milk production levels have increased dramatically in the past few years, causing many manure-storage sizing recommendations to underestimate actual requirements. If the storage is to be emptied once per year, a minimum of 12,000 gallons per cow per year should be expected. After the sketch with dimensions is created, stake out the area to ensure that the site is appropriate.

Expansion

No matter what size of dairy is being planned, try to reserve space for additional barns, feed storage, and so on, in case the need for them ever arises.

Barn Orientation

Barn orientation must be considered if freestalls are to be placed on outside walls (six-row or four-row tail-to-tail designs). Cows often refuse to use freestalls if the hot summer sun intrudes into the building (Figure 8-1). Roof extensions can be added to help protect cows on the outside rows of

FIGURE 8-1 Orienting a freestall barn incorrectly allows sun to penetrate the barn. Cows often will refuse to lie in sun-exposed freestalls. Adding a sun shield in this case helps shield the animals inside the barn.

stalls. The length of this eave extension should be about one third the height of the sidewall (i.e., a four-foot eave on a 12-foot sidewall). Wind, snow, and ice conditions on unshaded feed fences should be avoided during winter. Facing **open-sided barns** to the east or south to avoid the cold northern and western winter winds is recommended. Naturally ventilated facilities should be built far enough away from other buildings (Figure 8-2) and trees (Figure 8-3) to prevent them from obstructing air movement. Modern freestall barns are often separated by about 100 feet to avoid this problem. Placing a manure pit close to a freestall barn is not recommended, as manure smells may enter the barn.

FIGURE 8-2 Facilities that are too close together can retard airflow, making it more difficult to cool the animals.

FIGURE 8-3 A wood lot near a facility may decrease air movement, making it harder to keep cows cool and to remove stale air from the barn.

Prevailing Wind Speed and Direction

Seasonal **prevailing wind** speeds and directions are important from a cow-cooling standpoint in warm weather and cow-sheltering standpoint in cold weather. Check local wind maps to determine the most prevalent direction of summer breezes. Place barns so that the largest area of the structure faces the prevailing summer wind to enhance the cooling effect. Choosing the best building orientation on a site sometimes can involve a trade-off between wind direction (to maximize cooling) and **sun angle** (to protect cows from the sun on an outside row of stalls). Remember that the breezes crossing the barn to cool the animals will also carry odors and flies away from the farm site. Be careful not to place a dairy such that these nuisances are aimed at your or a neighbor's home.

Site Elevation

If possible, freestall barns should be built with a high **site elevation** to maximize the cooling effect of natural breezes blowing across the barn (Figure 8-4). "On a knoll, not in a hole," as the adage says. A site that provides a sufficient drop in elevation to support a **gravity-flow manure handling system** can result in decreased manure handling and lower equipment maintenance costs. Areas within a 100-year (or less) flood plain should be avoided.

Site Slope and Water Drainage

A site that is too flat or too hilly can add appreciably to the overall cost of a dairy. A slope of two- to six-percent will provide drainage without erosion. Good subsoil drainage will help prevent frost heave of foundations. The manure management system selected will influence the slope needed and the size of the manure storage facility needed. Flush systems and outside lots

FIGURE 8-4 Proper site selection involves placing the facility on a high elevation to facilitate animal cooling, natural ventilation, and proper drainage. Placing each structure sufficiently distant from other structures prevents obstruction of the air flow.

both require two- to three-percent slopes, which may influence the amount of soil that must be moved. A site on a high elevation is ideal if it does not require extensive rearrangement to provide sufficient space for current and future growth. A **topographic survey,** which shows the elevation, and the advice of local excavators can help you estimate the costs to prepare the site.

Groundwater Protection

Dairies must control both **nonpoint pollution sources** (such as cropland) and **point pollution sources** (such as direct discharge of milkhouse waste to a stream). Pollutants include manure, milkhouse waste, **silage leachate,** sewage, fuel spills, and so on. The local topography and distance from surface water (such as rivers, streams, lakes, and wetlands) should be considered. Building and site selection should support the diversion of clean stormwater away from—and manure-contaminated water into—a proper containment structure. Groundwater samples should be taken to verify that no current problems exist. Check local setback requirements, and place buildings at a sufficient distance from wells and sewage systems to avoid any contamination of groundwater.

Geological Homework

Proper site selection involves knowing what lies under the surface. A professional soil scientist can help you do the **geological homework** necessary to evaluate the appropriateness of the site. Soil borings should be taken to determine soil type; presence, type, and depth of rock formations; and depth of water table. Bedrock formations close to the surface can restrict installation of underground utilities and basement parlors, and may preclude use of earthen storages for water or waste.

Soil Type

Consider carefully the soil types of a potential site. Cropland should have good soil depth and nutrient holding capacity. Building sites require firm, stable subsoil with a low **shrink–swell potential.** Avoid sandy or porous soil types for building sites, since they have the potential for groundwater contamination. Lagoons and holding ponds require clay subsoil with low permeability.

Water

Each potential site should be checked to ensure that the water supply can support the long-term needs of the dairy herd being considered. Both quality and quantity of water are important. It is common for a dairy to use 100 to 150 gallons of fresh water per cow per day. Water should be tested to ensure that the quality is adequate for dairy cattle consumption. Well-driller logs for the local area and short-term pump tests may be used to approximate the expected availability of water on a site, but you may want to install a pump, and pump at the expected demand rate for several days to ascertain that sufficient water reserves exist.

Utilities

Determine how far the site is from access to electric and telephone utilities. **Three-phase power,** recommended for large dairies, can be expensive to bring to a new site.

Major Highways

A well-developed roadway is important to support the truck traffic associated with a large dairy. Time should be spent investigating local highway load restrictions. A state-maintained, hard-surfaced highway (with bridges permitting large trucks) should be easily accessible. Prompt snow removal is important. Avoid sites requiring excessive road construction and maintenance from the dairy to public roads.

Cropland Needs and Soil Nutrient Balance

A **nutrient budget** should be created to determine that sufficient land is available to maintain a nutrient balance (N-P-K) for the herd size being considered. Sufficient land should be under your control so that manure disposal will not cause nutrient buildup. In the Midwest, most feed delivered to the farm and manure returned to cropland have high moisture content, making them costly to haul. Current nitrogen-based standards often require close to two acres per cow for manure management. With phosphorous-based standards, which are likely to be implemented soon, this requirement could double in many areas. If land for manure application will not be owned, then it is advisable to establish long-term lease arrangements so that you have control over where, when, and how manure is applied.

Access to Cropland

Consider the distance and road access for hauling of feed and manure. If possible, select a site in the center of the cropland supporting the dairy.

Neighbor Relations

Environmental regulatory agencies often require minimum distances between dairy facilities and neighboring homes. Check existing zoning restrictions for any potential site. Notifying neighbors early in the planning process (and keeping them informed of your plans) will often help smooth the permit process and avoid hostile public meetings. Correct facility design, site selection, and proper management can help avoid **nuisance problems**—odor and flies, and road contamination issues—that annoy neighbors. During calm, humid periods, topography can funnel odors down drainage ways to distant locations, especially to residences located in valleys. Consult topographical maps to avoid this situation. Try to avoid sites near current or potential housing developments. Consider a buffer zone or a tree windbreak to shield the operation. A separation of a mile or more is recommended between large livestock operations and neighbors.

Environmental Regulations and Permitting

Each producer should have the objective to build an environmentally friendly facility. It is important to identify early in the process all permits needed and the regulating agencies (such as health departments, milk inspectors, designated manure-regulating agencies, local governments, etc.) who grant them. Start early and try to develop a strong, friendly working relationship with all of the regulatory agencies that have jurisdiction over the site.

Expansion on an Existing Site

For existing producers who want to modernize their operation, the most difficult decision may be whether to use an existing site or go to a new location. Lenders often encourage building on a new site, because the resulting facility may be more easily marketed. Producers often favor expanding on an existing site to save money and to allow a slower, phased growth pattern. Both objectives have merit; the decision between them should be based on the financial resources available and the ability of the present site to support the long-term growth needs of the operation. If financial resources are limited, building near and utilizing existing facilities makes good sense. If this alternative is considered, make sure that the existing site and the existing facilities are thoroughly evaluated. Do not overestimate the value of existing silos, manure pits, sheds, and so on. Consider the overall effect on the site of a plan with and without each existing structure. Building on a site that has limited growth potential should be avoided if possible.

Building a New Dairy

Site selection is in some ways simpler for a new dairy, because the producer need not consider its effects on an existing operation. It is often more costly, however, because additional improvements (such as roads, wells, and utilities) must be brought to the site. Working within the aforementioned constraints and evaluating the cost–benefit relationship of different options should help in selecting the best site.

If an existing producer decides to build a new facility some distance from the current facility, the use of the old facility should be considered in site selection. If animals will be housed at the old site, feed storage, feed delivery, and animal movement implications should be considered. The size, age, and condition of existing storage should be considered when determining where additional storage should be built. Housing far-off dry cows at a second site requires weekly movement of animals. If close-up animals will also be housed on a second site, daily movement of animals is often needed. The labor requirement associated with moving animals and feed, plus the ability to properly care for maternity animals, should be factored into the site-selection decision.

SUMMARY

Selecting the correct site for a dairy is the most fundamental—and often the most difficult—decision facing the dairy owner or manager. Producers have differing goals and circumstances, which add to the complexity of the decision. The ramifications can have a huge effect on cow comfort, labor efficiency, and the ability to support long-term growth. Organizing a team of experts who can help identify potential sites, site trade-offs, and growth strategies will often help improve the quality of the decision.

CHAPTER REVIEW

1. Describe a primary advantage and a primary disadvantage of phased expansion.
2. Identify five aspects of a site that must be evaluated when it is under consideration for dairy construction or expansion.
3. What is the purpose of a nutrient budget?
4. Identify the three ways in which a site's topography can affect its value.
5. Give two examples of an environmental regulatory agency.
6. Determine the site requirements of a 500-cow facility. What structures must be built, and how big should they be? What structural provisions must be made for feed storage, manure handling, hauling, and so forth? What site features would be ideal for this facility? Be sure to consider site size, water availability, utilities, access roads, and neighbors in your analysis. From your analysis, create a diagram of the site and a list of the facility's structural needs.
7. Imagine that the facility you described in your answer to the previous problem will be expanded to a 1,000-cow dairy operation over the next five years. Does the site you designed support this expansion? Modify your analysis, diagram, and facility list as necessary for the expanded size.

REFERENCES

Frank, G. G. (2002). Milk harvesting costs. Low cost parlor options (pp. 1–3) [CD-ROM]. Madison: University of Wisconsin—Extension.

Martin, J. C., III. (1998, January 28–30). Siting large dairy facilities. In *Fourth International Dairy Housing Conference,* (pp. 29–36) St. Louis. St. Joseph, MI: American Society of Agricultural Engineers.

Chapter 9

Milking Center Options

KEY TERMS

back-out flat-barn parlor
cluster spacing
coliform count (CC)
contact time
grouping milking routine
laboratory pasteurized count (LPC)
machine on-time
milking center
milking procedure
milking routine
preliminary incubation count (PI)
prep-lag time
prep time
sequential milking routine
somatic cell count (SCC)
standard plate count (SPC)
subway parlor
territorial milking routine
walk-through flat-barn parlor

Objectives

After completing the study of this chapter, you should be able to

- measure the milk quality and udder health of a herd.
- identify the advantages and disadvantages of minimal and full milking procedures.
- describe the physical differences between milking center types.
- understand how the structure of each type of center impacts milking procedures, milk quality, animal health, and overall productivity.
- evaluate an existing dairy to determine which type of milking center is most appropriate given the operation's goals, financial position, existing facilities, and operational style.

Cows must be milked, and there are many milking system options to choose from. The best **milking center** type differs among producers because of their site and situation. The same is true for the different features that are associated with each milking center option. Each of the available features may contribute to labor efficiency, cow comfort, or worker comfort, but the cost of each must be considered and evaluated based on the needs, paybacks, and financial position of the owner. For many producers considering expansion, cows are milked in the barn where they are stabled, whereas other producers already have an independent milking center where cows are taken to be milked. Expanding a herd is often easier for producers who have an existing milking center that can be more fully utilized, since only housing, feed, and manure capacity issues need to be addressed.

Milking Center Types

Milking centers vary greatly across the country. Herd size and climate have a bearing on a producer's choice. Producers with large herds normally select parallel, herringbone, or rotary parlors. Producers with smaller herds, and those making their first move away from milking and stabling their animals in the same barn, often choose lower-cost options. Switch milking, flat-barn parlors, swing parlors, used equipment, and the use of existing barns to house the milking center are common. These smaller operations often select parallel, herringbone, or auto-tandem (side opening) parlor types. With all herd sizes, the milking center choice should be based on the short- and long-term goals of the operator and the expected cost to harvest milk.

Switch Milking

Switch milking refers to using an existing barn and "switching" pens of animals into the facility to be milked. For instance, a producer with an existing 50-stall barn may build a new freestall barn to house cows in pens of 50 animals that can be taken to the old barn and milked. This option has a low capital cost because existing equipment and facilities are used, but the labor cost associated with milking and moving cows is often high. For smaller operations or producers with limited capital, this should be considered if the barn is located within reasonable walking distance of a proper building site that can support the operation's long-term expansion needs. Walking distances of over 500 feet have been used successfully in the past. Cows adapt well to the walk, but proper lighting, snow removal, and other worker comfort issues must be considered (Figure 9-1).

Flat-Barn Parlors

Flat-barn parlors (Figure 9-2) are normally built in existing buildings and are similar to switch milking in that cows are brought from a freestall barn or pasture to an existing barn to be milked. Rather than using all 50 stalls,

FIGURE 9-1 Correct siting of a new freestall barn is very important to the long-term growth of a dairy, as cows often must be walked outside during initial phases of an expansion.

FIGURE 9-2 Flat-barn parlors, with a limited useful life expectancy, can be a low-cost option and are often considered by dairies in transition.

some of the stalls would be used to milk cows, and other stalls would be removed to provide space for a holding area. For instance, eight stalls on each side of the barn could be left and four milker units used on each side. These eight units (four per side) would be used to milk half of the cows locked in the 16 stalls. When the first group of cows has been milked, the units are switched to the other eight cows and the just-milked cows removed from the barn, and eight more cows are brought from the holding area to be ready for the next shift of the milk machines. These flat barns are normally more labor efficient than switch milking in existing barns since the milking machine units do not need to be moved. To improve cow entry and exit from the milking-area, manure gutters are often filled or covered by gutter grates. To ensure good cow movement, gutter grates should be strong enough and correctly placed so the cow does not sense movement when weight is placed on them.

In **back-out flat-barn parlors** cows must back out of the stall after being milked; in **walk-through flat-barn parlors,** the cows can walk forward after being milked. Existing barns can often be converted to back-out flat-barn parlors for \$10,000 to \$15,000 if existing stalls can be used, whereas walk-through parlors may cost \$40,000 or more if new equipment is purchased. Worker comfort in these parlors is often better than in stanchion barns, in that the cows may stand at a higher elevation and less kneeling is required, but they are not so comfortable as a true pit parlor. Initial cost, worker comfort, worker safety, worker efficiency, ability to implement correct milking procedures, and the expected amount of time the parlor will be used should be considered when selecting among these options. Since worker comfort, safety, and efficiency are less satisfactory in a flat-barn parlor than in a pit parlor, the cost of a pit parlor should be ascertained before building a flat-barn parlor.

Swing Parlors

In a swing parlor, one set of milking machines is installed and used to milk both sides of a pit parlor (Figure 9-3). A double-8 parlor would have eight milking machine units. Eight cows on one side of the pit would be milked first and then the units swung to milk eight cows on the other side. This is the parlor of choice for producers who want to milk cows fast with a low investment per animal. Parlors are normally of the herringbone type, with varying **cluster spacing** depending on the animal angle selected for positioning the cows. These parlors are often built with a simple, inexpensive rump rail to restrain animals. To keep construction costs low, this rump rail often does not have manure splashguards, the platform floor does not have gutter grates, and so on, to reduce manure splattering. Therefore, these parlors are often considered dirtier places to work than conventional parlors. Milk lines are normally mounted above the milker units (referred to as a high-line), which may have an impact on vacuum levels and increase the incidence of mastitis. Since units from one side must be taken to the other side, a slow milking cow on one side may cause the cow on the other side to have her unit attached after her milk let-down response.

Herringbone Pit Parlors

Herringbone parlors get their name from the angle at which cows are positioned with respect to the pit wall. Normally a row of cows is positioned at a 45° angle on each side of the parlor pit. Milker units are applied to the udder from the side of the cow, and this configuration allows an arm-type take-off to be used. With American Holstein cows, herringbone parlors

FIGURE 9-3 Swing parlors are used in many parts of the world where a low-cost alternative and short milking shifts are desired.

require about 45 inches of parlor pit length per stall. This is less than auto-tandems and more than parallel parlors, and should be taken into account when building a large parlor. Since milking machines are applied from the cow's side, the animal can kick the operator more easily than she can in some other parlor types. Gutters and grates are normally used with herringbone parlors to catch manure and urine and to minimize manure splattering on the operator.

Parallel Pit Parlors

In parallel parlors, cows stand perpendicular to the pit wall, and milking machines are applied between the hind legs of the animal. Parallel stalls require only about 27 inches of pit length per stall and are normally chosen for large parlors to minimize walking distance. Parallel parlors are normally considered safer because stalls are built with a rear rail that prevents cows from kicking the operator. Since the milking machine must be attached between the legs of the animal, the operator must be careful to remove any manure from the milking machine claw to prevent contamination of the milk. No arm take-off has been designed for this parlor type, so a rope or chain take-off must be used. To minimize manure splattering, these parlors often have butt pans mounted on the rear of the stall to catch manure or urine.

In both parallel and herringbone parlors cows are loaded, milked, and released in batches. For instance, with a double-8, eight cows are loaded on a side, prepped, milked, and then released. This batch handling of cows can support a wide range of efficient milking procedures.

Milking Procedures and Routines Used in Parallel and Herringbone Parlors

The following terms refer to common milking parlor procedures.

Prep time—time taken to manually clean and dry the teat surface.

Contact time—the actual time spent manipulating or touching teats, which is the source of stimulation for oxytocin release.

Prep-lag time—time between the beginning of teat preparation and the application of the milking machine.

Milking procedures—the individual events (i.e., strip, pre-dip, wipe, attach) required to milk a single cow.

Milking routines—define how an individual milker or a group of milkers carry out a given milking procedure (minimal or full) over multiple cows. In parallel and herringbone parlors, there are three predominant milking routines: group, sequential, and territorial.

- **Grouping milking routine**—in a grouping routine, the operators perform all the individual tasks of the milking procedure on four or five cows. Once they have completed a group of cows, they move to the next group of available cows.
- **Sequential milking routine**—operators split up the individual tasks of the milking procedure among themselves, and work as a team
- **Territorial milking routine**—milkers are assigned units on both sides of the parlor and operate only the units assigned to them; they are not dependent on other milkers to perform specific tasks.

The two predominant milking procedures are minimal (strip or wipe, and attach) and full (pre-dip, strip, wipe, and attach). Milking procedures impact the number of cows per stall per hour in parallel, herringbone, and rotary parlors. In large parallel and herringbone parlors, cows per stall per hour were 5.2 when minimal milking procedures were used and 4.4 when full milking procedures were used. Cows per stall per hour declined from 5.8 to 5.3 when a full routine was used, compared to a minimal routine in rotary parlors (Armstrong, Gamroth, & Smith, 2001). In large parallel and herringbone parlors, milking procedures have a dramatic impact on the number of units one operator can handle. In 1997, Smith, Armstrong, and Gamroth published guidelines for the number of units that one operator could handle using a minimal and a full milking procedure. Using a full milking procedure, a milker could operate 10 units per side, compared to 17 units per side using a minimal milking procedure. These recommendations allowed four to six seconds to strip a cow, and four minutes to attach all the units on one side of the parlor.

In recent years, several milking management specialists have been recommending two to three squirts per teat (8–10 seconds) when stripping cows to increase stimulation and promote better milk letdown. Some of these management specialists believe that increasing the amount of stimulation reduces **machine on-times.** At present, a strong data set supporting this theory does not exist. An American Association of Bovine Practicioners (AABP) research update report by Johnson, Rapnicki, Stewart, Godden, and Barka (2002) indicated that milk flow rate decreased when cows that had been previously stripped were no longer stripped. If this change is implemented, producers will have to reduce the number of units one operator can manage per side (Table 9-1). The sequencing of the individual events of the milking procedure is critical. Rasmussen, Frimmer, Galton, and Peterson (1992) reported an ideal prep-lag time of 1 minute 18 second. Prep-lag times of 1–1.5 minute are generally accepted as optimal for all stages of lactation. Some of the advantages and disadvantages of minimal and full milking procedures are listed in Tables 9-2 and 9-3.

TABLE 9-1 Time (in seconds) required for individual events of the milking procedure.

	Procedure		
Event	**Minimal***	**Full**	**Full with 10-Sec Contact Times**
Strip	4–6	4–6	10
Pre-dip		6–8	6–8
Wipe	6–8	6–8	6–8
Attach	8–10	8–10	8–10
Total	12–18	24–32	30–36

*Strip or wipe, and attach

TABLE 9-2 Advantages and disadvantages of a minimal milking routine.

Advantages	Disadvantages
Successful when cows enter the milking parlor clean and dry	Compromises teat skin sanitation
Time required to milk the herd may be decreased (total milking time)	Machine on-time may be prolonged
Steady-state throughput increased	May require milkers to decide when extra cleaning of dirty teats is required; can cause lower milk quality and higher mastitis when compared to "full hygiene"

TABLE 9-3 Advantages and disadvantages of a full milking routine.

Advantages	Disadvantages
Maximizes teat sanitation and milk letdown	Uses four separate procedures (or can combine into two or three procedures)
Minimizes machine on-time	Results in lower cow throughput or higher labor cost
Maximum milk quality	Requires more milker training to maximize results

Three predominant milking routines are used in parallel and herringbone parlors (sequential, grouping, and territorial). These milking routines are presented in Figure 9-4. The use of territorial routines will reduce throughput 20–30 percent when compared to sequential routines (Smith, Armstrong, Gamroth, 1997). Grouping routines seem to be an alternative to sequential routines without sacrificing throughput. Sequential and grouping routines are demonstrated in Figure 9-5. In the planning of a parlor, the desired milking routine should be determined and procedures implemented that will be easy for employees to understand and follow.

Auto-Tandem Pit Parlors

Auto-tandem parlors are often referred to as side-openers because cows enter the milk stall from the side, off a lane that runs behind each row of stalls. In this parlor type, cows stand in a line, and the length of each stall must be as long as the cow being milked. Cows are loaded individually and are allowed to exit at random times. This is different than the parallel and herringbone parlors, which milk cows in groups. Since cows can exit when they are finished (they do not need to wait until the last cow in the group is finished), the number of cows milked per stall per hour is higher than with parallel or herringbone parlors. But because the stalls are longer, large parlors of this type are not recommended, which limits the maximum herd size to be milked with this type of equipment. Producers considering this option must consider the long-term growth potential of their operation and factor it into their buying decision. Since cows enter and exit at random times, it is difficult to efficiently implement certain recommended milking procedures.

Rotary Pit Parlors

In rotary parlors, cows enter from a fixed point, ride a rotating disk, and exit at a second fixed point (Figure 9-6). This technology is not new but has been regaining popularity lately with large operations. Cows seem to enjoy the ride and often fight to get on the rotating disk. Operators remain at fixed positions and cows pass by as the disk turns. One operator is usually

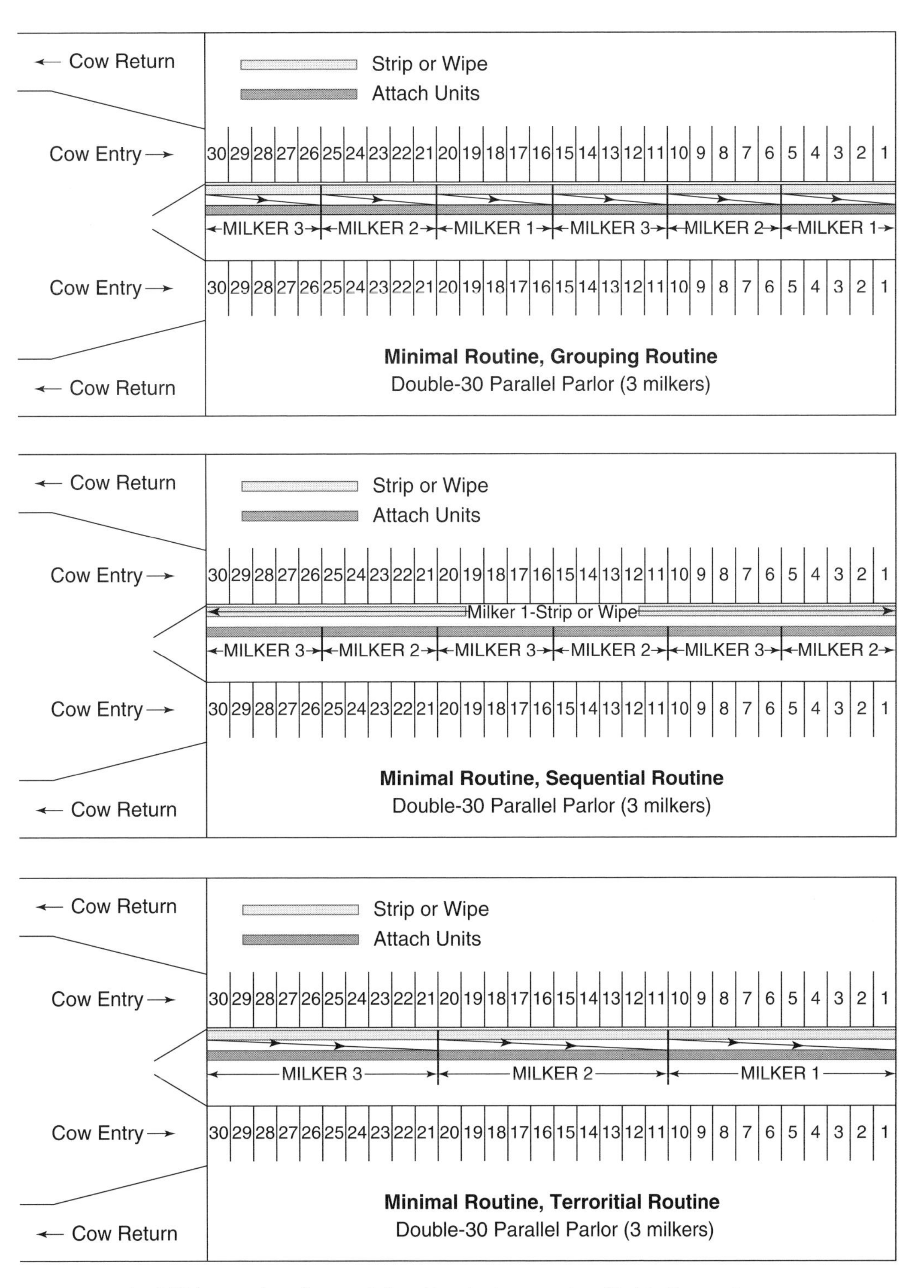

FIGURE 9-4 Milking routines for parallel and herringbone parlors (Smith, Harner, Armstrong, Fuhrman, Gamroth, Brouk, et al., 2003).

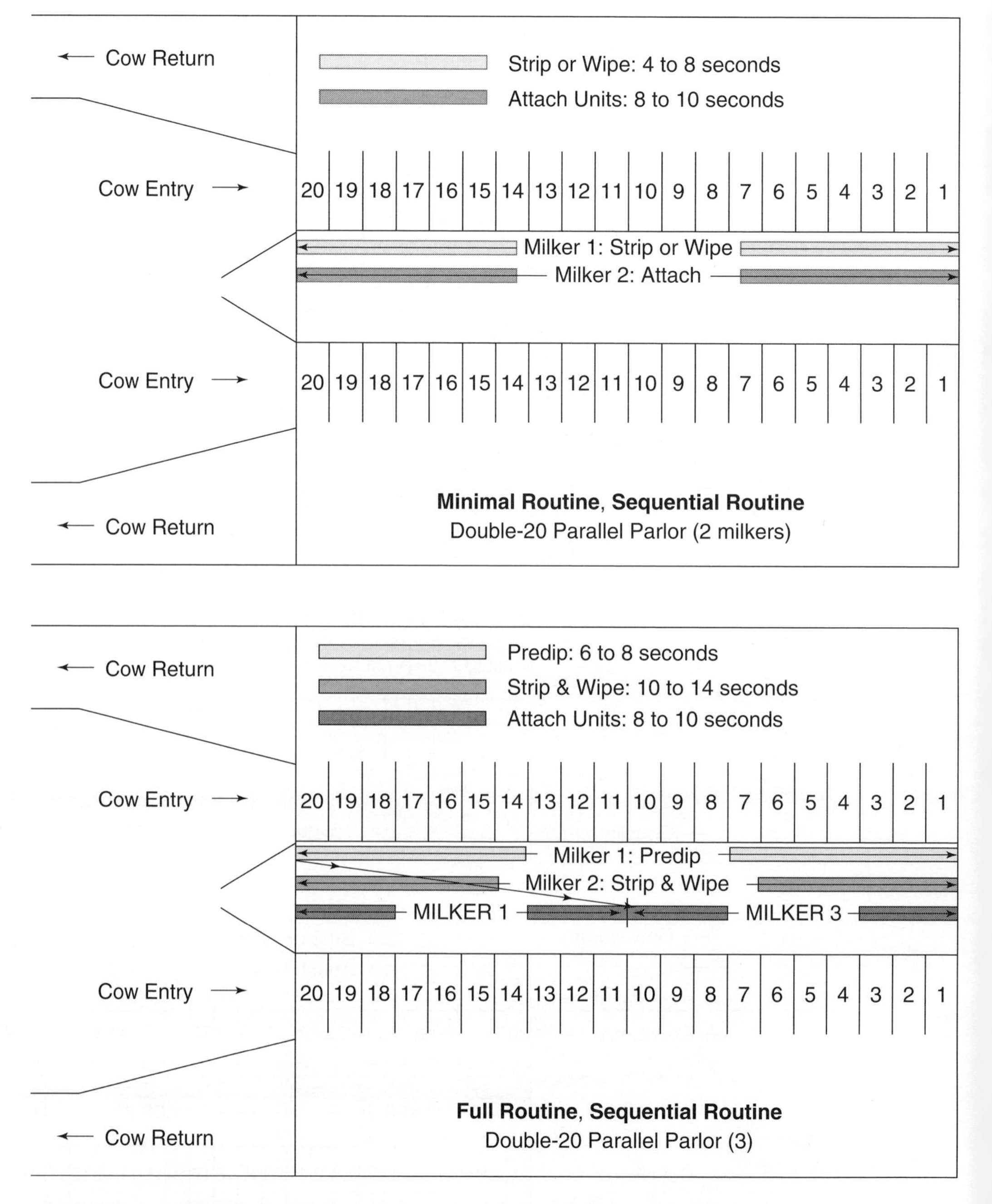

FIGURE 9-5 Sequential milking routines for double-20 parallel parlors using minimal or full milking procedures (Smith et al., 2003).

required for each function, and training of milk machine operators is simplified because each employee has only one function to perform. After the milking unit is attached, the cow moves past the operator who attached the claw. If a milking machine liner slips or a milking machine claw falls off, a worker must leave his position and go to the problem machine to adjust or reattach it. The capacity of rotary parlors is limited by the time it takes to

FIGURE 9-6 Rotary parlors are becoming very popular in large dairy operations.

load each animal onto the disk. If a 10-second average load time is achieved, a theoretical throughput of 360 cows per hour could be expected. (Since not every stall will get filled as the disk turns, some cows will require a second trip around in order to be completely milked; and since the disk also must be stopped in case of problems, it will never reach this theoretical limit.) Smith et al. (1997) recommend sizing these parlors assuming 11–12 seconds per stall to load cows, with 80 percent of the theoretical throughput expected. Figure 9-7 shows the typical full and minimal milking routines for a 72-stall rotary parlor (using four operators for a minimal routine, and five operators for a full routine). Notice how the available unit on-time changes.

Rotary parlors have a reputation for milking large numbers of cows per hour if a minimal milking procedure is used. Research by Smith et al. (2003) shows that their throughput, on a cows-per-worker-hour basis, is less than other parlor types when a full-prep milking procedure is used, because of the number of workers needed. Rotary parlors usually require a larger building than herringbone or parallel parlors with the same milking capacity, and have higher equipment costs. Furthermore, expansion of the parlor is difficult. It should also be noted that if an operator leaves a pit, a replacement worker is needed or the parlor must be stopped. Since operators continuously perform the same repetitive task, workers are often rotated among functions to avoid operator boredom.

Robotic Milkers

The newest way to milk cows is with a robotic milker (Figure 9-8). Developed in Europe, robotic milkers have been used on several hundred

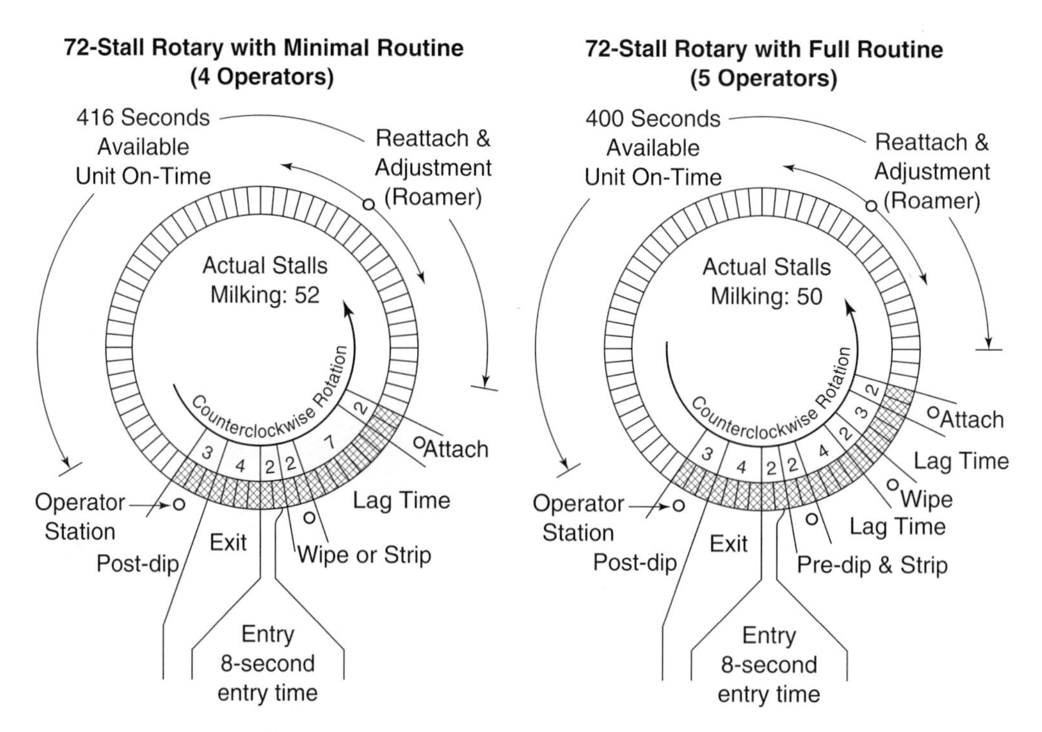

FIGURE 9-7 Minimal and full milking procedures in rotary milking parlors (VanBaale, Smith, Armstrong, & Harner, 2004).

FIGURE 9-8 Robotic milkers are new to most American producers.

farms for a number of years. They are currently being tested and sold, on a limited basis, in the United States and Canada. They hold a lot of promise for traditional producers who wish to increase herd size and milking frequency without adding hired labor. Price, milk quality issues, and current milk marketing regulations still need to be resolved.

Sizing a Milking System

Designing a dairy requires understanding the capacity of the milking system and the different factors that affect it. Throughput often increases by 8–10 percent when cows are switched from 2X to 3X milking, pre-dip milking hygiene reduces throughput by 15–20 percent, new parlors have 10–12 percent greater throughput than renovated parlors, and the average number of cows milked per operator per hour decreases as the number of operators increases. The expected throughput of parallel and herringbone parlors is normally between four and five turns (times filled per hour), depending on the operation's circumstances. For example, if a herd is milked three times per day using a double-12 parlor with clean cows, and 4.5 turns are expected, then all planning should be based on a 108 cows-per-hour throughput (24 stalls × 4.5 turns = 108 cph).

When selecting a milking system, both the initial and ongoing costs must be considered. You should compare the expected average annual total cost, cost per cow, or cost per hundredweight of milk shipped for each system. Be careful not to overestimate the useful life of milking equipment, which often must be replaced or upgraded in five to seven years because of wear and obsolescence.

Producers who are moving from traditional dairy operations to parlor systems tend to build parlors that are larger than needed. Producers often think that the extra parlor size will allow the herd to be milked quickly so they can move on to the other activities they are accustomed to doing. Owners in this situation should remember that if they do milk the cows, they are actually *earning* a wage equivalent to what a worker could be hired for. Often, after producers modernize their operations, the first employees they hire milk the cows. This is logical because milking is a repetitive task that employees can be easily trained to do. The producer who builds a parlor complex larger than needed therefore causes several problems: (1) the cost and payments are larger than needed; (2) extra cleaning materials, and so on, are required; and (3) it may be more difficult to find employees willing to work less than an eight-hour milking shift. Parlor size should be balanced with the other components of the dairy. Often a two-person crew is present during each milking even though one person can operate the parlor. This second person can decrease the milking shift length by moving groups of cows, policing freestalls, scraping manure, and giving the milker a break; plus he or she offers an element of security in case of an emergency. Selecting facility features and work schedules that integrate the needs of the total operation can in this way improve its overall efficiency.

Milking Center Components

A modern milking center consists of milking area, animal-holding pen, milking equipment area, milk storage or loading area, and often one or more offices. Each of these components should be located to minimize construction cost and labor utilization costs. Parlor complexes with equipment rooms and offices located on alternate sides of the milking parlor are often built to minimize the noise in offices, but they are more expensive than such facilities built on the end of the parlor building.

Operator Pits

Operator pits are normally eight feet wide between curbs. Larger parlors sometimes have wedge-shaped operator pits narrowing from about 10 feet wide at the breezeway to six feet at the other end. This arrangement offers two advantages: it gives people standing at the breezeway end a better view of the parlor, and it provides more space for cows to exit at the far end. The cow platform height should be sized for the type of people that will milk in the parlor, with 38–42 inches above the floor of the operator pit being typical. If in doubt, it is better to make the platform too high, because mats, and so forth, can always be used to raise the operator. Movable floors have become quite common in large parlors because of the extra comfort given to the milker (Figure 9-9). In parallel parlors, the operator pit floor should have about two inches of side slope from the center of the pit to the pit wall and 1 percent slope for the length of the parlor.

Subway (or basement) **parlors** (Figure 9-10), in which the milklines and some milking equipment are placed below the parlor pit or cow platform, increase the initial building cost but decrease parlor noise, decrease equipment cleaning needs, allow more flexibility in the placement of milklines and receiver jars, and allow for easier maintenance of milking equipment in large parlors with long operating schedules.

Parlors with rapid-exit stalls allow all cows in a group to move directly out of the milking stalls through individual gates or a raised barrier in front of the cows. Exit time is decreased and normally can be cost-justified on double-10 or larger parlors.

The amount and type of automation selected depends on the parlor size, labor availability, capital resources available, and individual preference. For most parlors, automatic detachers should be considered because of the labor efficiency gains and the standardization of milk claw removal that they provide. Other mechanization to consider includes crowd gates, electronic animal identification, sort gates, and indexing (which allows the front rail of the milk stall to adjust to animal sizes).

FIGURE 9-9 Movable parlor floors are made to adjust to worker height and thereby increase worker comfort.

Holding Pens

A holding area is an integral part of all milking center systems (except switch milking). It should be large enough to hold the largest pen of cows to be milked, not only for the current expansion, but any future expansions contemplated. A minimum of 15 square feet of floor space per animal is recommended. To decrease the total milking shift length, most producers have another group of cows brought to the holding area before the first group has been finished. The milk machine operator can then finish one group and immediately start the next group. To accomplish this, the holding area capacity should be 125 percent of the size needed for one group. A series of gates and drover lanes must be designed so that animals from the two groups are not mixed up during milking.

The holding area should be sloped up to the parlor floor to encourage cows to stand facing the parlor; a slope of three to five percent is preferred. Adding a crowd gate to encourage animals to move into the parlor is also

FIGURE 9-10 Large herringbone and parallel parlors often have basements under the parlor pit floor. This helps keep equipment clean and simplifies maintenance.

recommended. People-passes, which allow people to move through a fence opening, are not recommended in the holding area because cows may stick their heads in them and get jammed by a moving crowd gate. If an opening is desired into the holding area, a gate should cover it. Ventilation is extremely important; proper summer ventilation and use of fans and sprinklers can help minimize the stress on cows waiting to be milked. It has become common to cover holding pen floors with rubber alley mats to increase cow comfort.

Exit Lanes

In parlors with 15 or fewer stalls on a side, a three-foot-wide exit lane leading away from the parlor is sufficient. For large parlors a six-foot lane is suggested. In cold climates this lane may freeze, causing cows to slip and fall when exiting the parlor. Producers have addressed this problem by installing

- heated exit lane floors;
- wide exit lanes (Figure 9-11) with skid steer access to remove frozen materials;
- gated exit lane fences that can be opened to remove frozen materials; or
- a warm holding area, which prevents freezing.

The number and location of exit lanes varies by parlor type and the animal handling system being considered. Parlors in which cows are milked on two sides of an access lane (for example, the pit of a pit parlor) often

FIGURE 9-11 Wide return lanes beside the holding area allow equipment to remove frozen manure in the winter.

have an exit lane from each side. This allows cows to exit the parlor without crossing over the access lane. If electronic sort gates are being considered, a parlor that allows cows to cross over the parlor pit before exiting, or allows cows from the two exit lanes to merge later, will support the need for one sort gate to select animals and one holding area for selected animals.

Foot Baths

Hoof care is a critical part of herd management. Foot baths in exit lanes have been used very effectively to reduce the incidence of foot-related problems. Placing two foot baths, one with clear water and a second with medicated solution, at least ten feet apart allows the first foot bath to clean the foot. The space between, which allows cows to defecate before reaching the second bath, makes the second foot bath more effective. The foot bath should be long enough to ensure that every foot of every cow is covered every time it is being used. Foot baths should be a minimum of eight feet long and support a minimum liquid depth of four inches throughout the length of the bath. Remember to have a water source easily accessible, and design foot baths that are easily cleaned (Figure 9-12).

Offices

To effectively monitor a modern dairy operation, good records are a must. A clean, comfortable working environment improves worker efficiency. A location near related activities minimizes the time needed for employees to access information and allows the manager to monitor the operation. Modern dairy equipment and the computers used by them should be housed in warm, clean areas. Rugs and foot-washing equipment should be strategically placed at entrances to reduce the amount of filth deposited in office and equipment areas. If resources are limited, office space may be minimized during early stages of an expansion, but long-term plans should allow for additions later.

FIGURE 9-12 Foot baths should be designed for easy cleaning.

Use of Existing Buildings to House Milking Centers

Producers considering the use of an old building to house a new milking center (Figure 9-13) should be very careful. Both options—placing the milking center in a new building or using an existing one—should be thoroughly evaluated and cost-compared. The usefulness, labor efficiency, and resale value of each option should be considered. Use of an old dairy barn to house a milking center can be an advantage if the barn is in good repair and is located on a site that supports the long-term needs of the dairy. Traditional producers can often install a pit parlor milking center in an old dairy barn for $50,000–$90,000, whereas the same parlor may cost three to four times that if placed in a new building at a new site. Use of an existing structure often is less expensive because the shell, electrical system, water system, and roads are already in place, but remodeling is often very costly and is notorious for cost overruns because of unforeseen difficulties. If an old barn is used, how will it be ventilated (Figure 9-14), and at what cost? When considering constructing a new building to house a milking center, a new site away from existing structures is almost always a good decision.

Milking Procedure Implications

There are many different ways to milk cows. Experts agree that the key element is to be consistent in whatever procedure you select. Circumstances vary by farm, allowing some herds to use procedures that would not work in other herds. Sometimes circumstances change, requiring procedural changes. Understand the basics of milk let-down, mastitis transmission, and so on, and their implication on milking procedures for each system being considered. Investigate the effects on labor usage and comfort with the

FIGURE 9-13 Housing a pit parlor in an existing dairy barn can be a very cost-effective decision. (Note how part of the barn is used as a holding area, and the old stanchions are used for animal handling.)

FIGURE 9-14 Existing barns can often be converted to contain pit parlors and holding areas, but ventilation concerns are paramount.

different milking center options considered. Look at the work-routing implications of different milking systems, as in the following examples, before making your decision.

- Cows are milked in batches with parallel and herringbone parlors, allowing great flexibility in procedures.
- Milkers stand and perform functions (apply pre-dip, wipe udder, attach milker claw, etc.) as cows move by on rotary parlors; this means that adding an additional function often means adding another worker.
- Walk-through flat parlors with one stall per milker unit, and auto-tandem parlors, process cows individually. These parlors require milkers to move about the parlor chasing the next cow that is ready. As a result, full-prep procedures are difficult, and worker efficiency decreases.

Efficiency Implications

When the different milking center designs are being considered, the issues of air-flow, cow-flow, and people-flow are critical. Proper ventilation of both parlor and holding area is important to reduce stress on cows and workers. Remember to think through both warm- and cold-weather implications. Concentrate on cow-flow issues relating to getting cows into and away from the parlor multiple times per day, plus any sorting and handling needed. Human access lanes, people-passes in gates and fences, and quick-opening gate latches all improve labor efficiency and reduce worker frustration.

Milking System Performance

A recent survey of Wisconsin producers who increased herd size by at least 40 percent from 1994 to 1998 showed performance values for different systems selected (Wagner, Palmer, Bewley, and Jackson-Smith, 2001). Table 9-4 shows that the smaller, less-expensive milking facilities selected by producers with smaller herds do not achieve the labor efficiency that pit parlors in new barns achieve. The average herd size for these groups ranged from 65–411 cows, with 7–20 miker units costing $4,191–15,832 per milking unit, and they can handle 21–43 cows per milker hour, where milkers include both people actually milking and those moving cows.

When only the producers who selected pit parlors are analyzed (Table 9-5), parallel parlors appear to be the most common pit parlor type with Wisconsin herd owners that expanded. Producers selecting herringbone parlors reported paying less per stall for their milking center, which indicates that a large percentage of these are using old equipment or placing

TABLE 9-4 Milking performance of different system types.

	Stall Barn with Pipeline	Flat Parlor in Old Barn	Pit Parlor in Old Barn	Pit Parlor in New Building
Herds, no.	65	52	73	107
1998 mean herd size	117	157	212	411
1998 RHA milk, lbs	20,684[c]	21,397[bc]	22,207[ab]	23,073[a]
Linear somatic cell score	3.02[a]	2.97[ab]	2.86[ab]	2.78[b]
Number of milking units	7[d]	9[c]	14[b]	20[a]
Number of milkers	2.4[a]	2.2[ab]	2.0[b]	2.1[ab]
Cows per hour	47[c]	55[bc]	61[b]	83[a]
Cows per milker hour	21[d]	27[c]	34[b]	43[a]
Time spent milking*	3.03[c]	3.92[ab]	3.78[b]	4.12[a]
Physical comfort of milker*	2.45[c]	3.83[b]	4.10[ab]	4.32[a]
Milk quality*	3.28[b]	3.75[a]	3.66[a]	3.70[a]
Safety of operator*	3.31[c]	3.40[c]	4.01[b]	4.38[a]
Cost per milking unit	$4,191[b]	$4,954[b]	$6,500[b]	$15,832[a]

*Average satisfaction reported on a scale of 1 (very dissatisfied) to 5 (very satisfied)

TABLE 9-5 Milking performance of different pit parlors.

	Auto-Tandem	Herringbone	Parallel
Herds, no.	7	67	104
1998 median herd size	190	275	245
1998 RHA milk, lbs	22,146	22,715	22,721
Linear somatic cell score	2.76	2.81	2.82
Number of milking units	10[b]	17[a]	18[a]
Number of milkers	2.1	2.0	2.2
Cows per hour	60	71	75
Cows per milker hour	30	38	40
Turns per hour	6.20[a]	4.13[b]	4.44[b]
Time spent milking*	4.00	3.89	4.02
Physical comfort of milker*	4.71	4.05	4.30
Milk quality*	3.71	3.74	3.62
Safety of operator*	4.14	3.85	3.88
Cost per milking unit	$17,268[a]	$8,944[b]	$13,201[a]

*Average satisfaction reported on a scale of 1 (very dissatisfied) to 5 (very satisfied)

the parlors in existing buildings. Performance and satisfaction for parallel and herringbone parlors are very similar.

Note: the superscripts shown in Tables 9-4 and 9-5 indicate whether the means shown within the table are significantly different ($P < 0.05$) (Wagner et al., 2001). If two mean values are significantly different, they will have a different superscript, and the largest value will have the superscript *a.* (An *ab* superscript is not significantly different from an *a* or *b* value, but is significantly different from a *c* value, etc.). For example, the average number of milking units for pit parlors in new buildings was 20, which was significantly larger than the average number of units (14) for pit parlors in old barns. Larger values are often good, but for some, like linear SCS (Somatic Cell Score), the smaller number is better.

Parlor Performance

It is suggested that producers understand the principles of monitoring parlor performance and use them when selecting the size and type of parlor to build. Planning for labor needs and using these plans in the development of your feasibility studies is a very important part of the process. When a facility (or facility change) is being planned, it is important to understand the roles of both the manager and the employees in the production of high-quality milk, and to select design features that support the accomplishment of this task in an efficient manner.

Measuring Parlor Performance

Everything revolves around the parlor. Because parlors are fixed assets, increasing their use increases profits. Milking cows 21–22 hours a day, depending on the time required for properly washing the system, makes the best use of this asset. Milking parlor performance has been evaluated by time and motion studies (Armstrong & Quick, 1986) to measure steady-state throughput (cph). Steady-state throughput does not include time for cleaning the milking system, maintenance of equipment, effects of group changing, and milking hospital strings. These efficiency measurement studies also allow us to look at the effects of different management variables on milking parlor performance. Some typical efficiency measurements are cows per hour (cph), the total number of cows milked in one hour; cows per labor hour (cplh), cph divided by the total number of milkers; milk per hour (mph), the total amount of milk harvested in one hour; milk per labor hour (mplh), mph divided by total number of milkers; and turns per hour (tph), the number of times cows enter and exit a parlor in one hour (also called parlor throughput).

Understanding these milking parlor efficiency measures can be very beneficial in making the correct parlor buying decision. When parlor performance is evaluated, parlor throughput can be further broken down into several individual time measurements, such as the following.

- Time from exit of the previous group until the first cow is touched (only if forestripping before pre-dipping)
- Time from exit of the previous group until the first cow is pre-dipped
- Time from pre-dipping to drying (check minimal "kill time")
- Time from exit of the previous group until the first unit is attached
- Time from exit of the previous group until all units are attached
- Time from exit of the previous group until all units are detached
- Time from when all units are detached until exit again

Measuring Effectiveness of Routines and Procedures

Regardless of the milking procedure and routine chosen, employees will be more receptive to and effective at performing procedures if they have a role in developing them. Employers can learn from employees, and incorporating workers in decisions that affect their work improves morale and the working environment in general. Employee input is crucial! The easier a job is to understand, the easier it is to manage. Keeping the routine as simple as possible and ensuring that employees perform equal amounts of work will minimize employee turnover and improve labor efficiency. The challenge for many dairies is to motivate milkers to properly clean teats prior to attaching units. Conducting milker meetings to clearly explain the procedures expected in the parlor, and why each step is important, has proven successful for numerous dairies. Milking procedures should be written (in the language of choice) and given to all milkers prior to performing the procedure. It is also beneficial to have procedures posted on the wall in the parlor for everyone to see.

There are numerous measures of milk quality that management must clearly understand. Facilities should be designed, goals established, and procedures implemented that facilitate acceptable levels for each of these measures. Table 9-6 explains typical milk quality, udder health, and general clean-in-place (CIP) sanitation measurements (**standard plate count, laboratory pasteurized count, coliform count, preliminary incubation count, somatic cell count,** clinical mastitis, teat and teat-end condition, added water, antimicrobial drug residues, and sediment) and the influence on these measures of both management and employees (adapted from VanBaale, Fredell, Bosch, Reid, & Sigurdson, 2001).

If milk quality and udder health goals are being met, then the milking procedure and equipment being used is most likely acceptable. Additionally, teat and teat-end condition scoring should be done on a regular basis to evaluate the health of the udder.

TABLE 9-6 Milk quality, udder health, and sanitation measures.

Standard plate count (SPC)	The SPC is the total quantity of viable bacteria in a millimeter (ml) of milk. The SPC is a reflection of the sanitation used in milking cows, the effectiveness of system cleaning, and the proper cooling of milk.
Employee influence	The manner in which cows are prepared for milking.
Management influence	The quality of water and the ability of the water heater to produce water of the appropriate temperature.
Laboratory pasteurized count (LPC)	The LPC is a measure of bacteria that survive pasteurization. This group of bacteria has an influence on the flavor and shelf life of dairy products. The general sanitation of the CIP system and the condition of the rubberware can contribute to a high LPC.
Employee influence	The manner in which cows are prepared for milking, as well as attention to the condition of rubber goods and the wash-up.
Management influence	The bacterial quality of the wash water and the choice of detergents and sanitizers.
Coliform count (CC)	The CC measure reflects the extent of fecal bacteria exposure of milk. Coliform bacteria can enter milk as a result of milking dirty, wet cows or may result from coliform growth within the milking system.
Employee influence	Employee hygienic practices have substantial control over the CC. The milking of clean and dry udders will limit exposure.
Management influence	CC problems may be associated with a poor CIP system.
Preliminary incubation count (PI)	The PI count is a measure of bacteria that will grow well at refrigerator temperatures. The PI is controlled by strict cow sanitation and excellent system cleaning.
Employee influence	Udder preparation and sanitation have a positive effect on the PI.
Management influence	The efficacy of the CIP washing system.
Somatic cell count (SCC)	The SCC on bulk tank milk and individual cow milk is a direct measure of the severity of mastitis (udder infection). The incidence and prevalence of the disease in the dairy is subject to a variety of factors. In general, the SCC reflects a subclinical or nonvisible form of the disease.
Employee influence	The manner in which the cows are milked can have a significant influence on the rate of new infections.
Management influence	The condition of the cow bedding environment and the commingling of chronically infected cows with noninfected cows.
Clinical mastitis	A proportion of mastitis infections become severe enough to become clinical. The clinical signs include changes in milk appearance and may include signs of disease in the animal as well. Milk from cows with clinical mastitis cannot legally be included in the commercial supply. It is the milker's responsibility to assure that the disease is detected early and the milk is diverted for discard or noncommercial use.
Employee influence	The employee has an influence on the manner in which cows with clinical mastitis are managed. Effective mitigation of the disease depends on prompt detection and management. A delay of 8–12 hours

TABLE 9-6 (Continued)

	can result in the incorporation of poor quality milk into the commercial milk and may result in greater disease costs.
Management influence	Type of teat dip used, the condition of the cow bedding environment, and the commingling of chronically infected cows with noninfected cows.
Teat and teat-end condition	The condition of teats is a direct reflection of the cow's environment, the use of teat dips, equipment settings, functionality, and upkeep. In addition, milking procedures and how well they are being followed impact teat and teat-end condition.
Employee influence	Adequately covering all of the teats, performing basic equipment checks and maintenance, and following a well-designed milking SOP (standard operating procedure) to the letter.
Management influence	Type of teat dip used, the condition of the cow bedding environment, implementing a well designed milk procedure, and maintaining properly functioning equipment in the milking parlor.
Added water	Milk is routinely tested for added water, using the freezing-point test. Dishonest producers sometimes add water to milk in order to increase the volume. Water may be added accidentally to milk by failure to drain the milking system fully before the milking begins.
Employee influence	During wash-up and sanitation of the milking system, the employee can ensure that all excess water is drained from the system. In the case of farms that have a several-hour period between milkings, standing water in the system may also be associated with elevated bacterial counts.
Antimicrobial drug residues	Most antimicrobial drug residues are not tolerated in milk; a few have legal tolerances, although the levels are extremely low. The type of drug and the manner of its application can greatly influence the potential for milk residues. Regulatory scrutiny has made dairy producers increasingly accountable for eliminating drug residue in milk.
Employee influence	Dairy farm management that instructs the employee to medicate cows for specific problems also must expect that the employee will be able to withhold bad milk from the commercial supply. This employee must know which cows are medicated and how long the milk is to be withheld. Some dairy farm employees are instructed in the use and interpretation of milk residue tests.
Sediment	The sediment in milk is a measure of the general filthiness of cows. This fine debris moves through the farm milk filter and is detected by the milk processor. High sediments may be associated with higher bacteria counts. However, some bedding materials, like river sand, may contain very fine particles that are measured in the sediment evaluation.
Employee influence	The general methods for cow and udder preparation will affect the amount of sediment in the milk.

SUMMARY

Selecting the correct milking center is not easy. It can be a very time consuming and confusing because of the many options available. Many different features exist that can improve operation efficiency, but features must be selected based on their relative value and the total resources available. When considering milking center designs, air-flow, cow-flow, and people-flow issues are critical. Proper ventilation of both the parlor and holding area is important to reduce stress on cows and workers. Remember to think through both warm- and cold-weather implications.

Concentrate on cow-flow issues relating to getting cows into and away from the parlor multiple times per day, plus any sorting and handling procedures needed. Provide human-access lanes, people-passes in gates and fences, and quick-opening gate latches to improve labor efficiency and reduce worker frustration. Remember that each feature should be evaluated not only on its own merit but also on how it complements other decisions with respect to both the initial capital cost and long-term labor implications. Since the system selected must be used for a long time, a thorough evaluation of available options and strategies is required.

Tour existing farms and talk to producers and the milkers who use the equipment every day. Go to farms during milking, and watch cows and milkers to determine the comfort of each. Watch milking procedures and ask about somatic cell count and mastitis. If possible, actually milk in different types of milking centers. Keep an open mind and take notes about the advantages and disadvantages of each system. Think about the implications associated with each system, and try to envision things that could happen to require you to change your expected milking procedure. For example, what are your options and costs if you select a milking center design thinking that no udder preparation will be needed, and then you encounter a severe mastitis problem that requires going to a full-prep procedure? Select a system that offers the flexibility needed, falls within your budget, and complements your current and long-term goals.

CHAPTER REVIEW

1. List the advantages and disadvantages of minimal milking procedures. List the advantages and disadvantages of full milking procedures.
2. What is one cause of mastitis? How can milking procedures be changed to reduce incidence of mastitis?
3. Using the facility descriptions in this chapter, create diagrams for switch, flat-barn, swing, parallel, herringbone, auto-tandem, and rotary parlors. Be sure that each diagram identifies cattle entry and exit points, placement of milking equipment, rump rails, and other pertinent features.

4. For each of the parlor types in question 3, explain how the structure of the parlor affects milking routines and animal health.
5. Imagine a dairy operation with a double-12 parallel pit parlor that is being switched from minimal to full milking procedures. On average, what is the dairy's new throughput? How much did the throughput decrease?
6. What are the five ways to measure parlor performance?
7. What are five key design features of an animal holding pen?

REFERENCES

Armstrong, D. V., & Quick, A. J. (1986). Time and motion to measure milk parlor performance. *Journal of Dairy Science, 69* (4), 1169–1177.

Armstrong, D. V., Gamroth, M. J., & Smith, J. F. (2001). Milking parlor performance. In *Proceedings of the 5th Western Dairy Management Conference,* Las Vegas, NV (pp. 7–12). Manhattan: Kansas State University Agricultural Experiment Station and Cooperative Extension Service.

Johnson, A. P., Rapnicki, P., Stewart, S. C., Godden, S. M., & Barka, N. (2002). A field trial to evaluate the effects of fore-stripping on milk-flow rates. Abstract in *Proceeding of the American Association of Bovine Practicioners 35th Annual Convention,* Madison, WI. Rome, GA: American Association of Bovine Practitioners.

Rasmussen, M. D., Frimmer, E. S., Galton, D. M., & Peterson, L. G. (1992). Influence of premilking teat preparation and attachment delay on milk yield and milking performance. *Journal of Dairy Science, 75,* 2131.

Smith, J. F., Armstrong, D. V., & Gamroth, M. J. (1997). Labor management considerations in selecting milking parlor type and size. In *Proceedings of the Western Dairy Management Conference,* Las Vegas, NV (pp. 43–49). Manhattan: Kansas State University Agricultural Experiment Station and Cooperative Extension Service.

Smith, J. F., Harner, J. P., Armstrong, D. V., Fuhrman, T., Gamroth, M., Brouk, M. J., et al. (2003). Selecting and managing your milking facility. In *Proceedings of the 6th Western Dairy Management Conference,* March 12–14, Reno, NV. Manhattan: Kansas State University Agricultural Experiment Station and Cooperative Extension Service.

Stewart, S., Godden, S., Rapnicki, P., Reid, D., Johnson, A., & Eicker, S. (2002). Effects of automatic cluster remover settings on average milking duration, milk flow, and milk yield. *Journal of Dairy Science, 85,* 818–823.

VanBaale, M. J., Fredell, D., Bosch, J., Reid, D., & Sigurdson, C. G. (2001). *Milking parlor management, quality milk production, from harvest to home.* St. Paul, MN: Ecolab.

VanBaale, M. J., & Smith, J. F. (2004). *Parlor management for large herds.* Charlotte, NC: National Mastitis Council. [Paper available upon request: vanbaale@ag.arizona.edu].

VanBaale, M. J., Smith, J. F., Armstrong, D. V., & Harner, J. P., III (2004). *Making decisions regarding the balance between milk quality, udder health, and parlor throughput.* [Technical paper]. Manhattan: Kansas State University Agricultural Experiment Station and Cooperative Extension Service.

Wagner, A., Palmer, R. W., Bewley, J., & Jackson-Smith, D. B. (2001). Producer satisfaction, efficiency and investment cost factors of different milking systems. *Journal of Dairy Science, 84,* 1890–1898.

Chapter 10

Feeding the Dairy Herd

KEY TERMS

bunker
commodity storage
feed shrinkage
feeding waste
fermentation waste
forage dry matter (DM)
harvest waste
horizontal silo
silage bag
stack
upright bin

Objectives

After completing the study of this chapter, you should be able to

- identify the key components of a feeding system.
- explain the advantages and disadvantages of various feed storage facilities for forage, grain, and supplements.
- explain the advantages and disadvantages of various TMR mixers.
- define a feeding program, choose a feed storage method, and select an appropriate TMR mixer for an existing dairy that will accommodate its short- and long-term goals.

One of the basic questions that must be asked in planning an expansion is how animals will be fed and manure disposed of. In northern climates, both feed and manure contain a large percentage of moisture, which makes it costly to transport them. The feeding program and feed, storage system selected should be based on the long-term availability and cost of feed, and feed and manure handling costs. Selecting a feeding system is not as simple as selecting the right total mixed rations (TMR) mixer. It starts with determining the herd feeding program. Considerations include feed acquisition planning, cropping enterprise evaluation, and feed storage, loading, mixing, and delivery implications.

Defining Your Feeding Program

Defining the best feeding program for a herd involves knowing the supply and costs of each ingredient. The feeding program for an operation may change between phases of the operation's development because of changes in the total needs of the herd and in the local supply of individual forages. Feeding trials generally demonstrate similar milk production from cows fed diets based on either corn or alfalfa silage. Use of a single forage source may require better herd management—or, at least, different operating procedures—to prevent adverse effects on cow health or milk production. In the Midwest, use of at least one-third each of alfalfa and corn silage is recommended to reduce risk of crop loss, spread labor requirements, and make better use of on-farm nutrients. As herd size increases, the percentage of corn silage in the diet normally increases because of its yield, harvesting, packing, and palatability advantages. Each dairy manager must consider local economics to define the operation's feeding program, and then plan acquisitions and storage requirements accordingly.

Determining Feed Storage Needs

Feed storage for forage, grain, and concentrates must be determined. Forage, the largest component of dairy rations, can be stored in **upright bins, horizontal silos** (**bunkers** or trenches), **stacks, silage bags,** or dry hay storage. Upright silos are the logical choice for small dairies, but they have limited value with large dairies because of the time required to fill and unload them (Figure 10-1). Bag storage can result in high feed quality, low feed wastage, and low annual cost, but bag disposal and animal and weather damage must be considered (Figure 10-2). Bags are often the logical choice of expanding dairies during their transition phases because of their minimal capital expenditure, and location and capacity flexibility. Flat silage stacks using crushed rock, blacktop, or concrete bases provide satisfactory results if properly managed (Figures 10-3, 10-4, and 10-5). Bunkers increase capacity, improve packing ability, and improve safety over silage stacks

FIGURE 10-1 As herd sizes increase, unloading of tower silos can delay loading of TMR mixers. Building a buck-wall and having silage ready to load can prolong the useful life of these structures.

FIGURE 10-2 Silo bags offer a very cost-effective feed storage alternative. They allow the operator flexibility in storage and feeding of different feeds. Feed quality is often excellent, but disposal of the used bag can be a challenge.

(Figures 10-6, 10-7, and 10-8). They can result in lower silage loss than improperly built and poorly maintained stacks. Grain and concentrates are normally stored in upright bins or flat-floored **commodity storage** areas. Bins with large augers can unload feed rapidly, minimize waste, and lead to accurate loading because of their precise cut-off controls (Figure 10-3). Having to take the mix wagon to the bin to load ingredients is a

FIGURE 10-3 Large modern dairies often use flat pads for forage pile storage, and upright storage bins. Upright bins cut waste and increase accuracy of feed loading. This photo shows a blacktop-based silage pad before silage piles were made.

FIGURE 10-4 Small piles have been successfully used to store feed on well-drained bases. This practice can cut capital expenses during the initial phases of an expansion.

FIGURE 10-5 Bunker silos should be sized and managed to remove a sufficient amount of feed from the face each day to prevent spoilage and animal health problems. This photo shows a feed face that was too large for the herd size, and silage damage can be seen at the top of the pile.

FIGURE 10-6 Large storage spaces, which allow flexibility in feed storage, are essential for large dairies. Silage should be carefully covered to minimize feed wastage. Operators removing feed from tall feed faces must be very careful to avoid face collapses.

FIGURE 10-7 Silage bunkers or flat pads support rapid loading of TMR mixers. Adding sidewalls can increase storage capacity and reduce packing risks. Losses can be reduced dramatically if silage is properly covered.

disadvantage. Flat grain and commodity storage allows quick loading of ingredients into mixers but must be carefully managed to minimize **feed shrinkage** and to ensure that accurate amounts of each feed are loaded into mixers (Figure 10-9). The number and size of feed storage units are determined by the total capacity needed by the operation at each phase of its growth, and the number of ingredients in the rations. Individual units should be sized to ensure a proper removal rate based on herd size and feeding rate. Multiple silos increase handling and management flexibility.

FIGURE 10-8 Small bunkers can be used very effectively by small- to medium-sized dairies.

FIGURE 10-9 Commodity sheds that house feed and bedding materials facilitate loading of TMR mixers, but feed losses caused by wind, birds, and weather must be considered.

Feed Mixing and Delivery Options

Total mixed rations, in which all forages, grains, and supplements are blended together into a homogeneous mixture prior to feeding, are used in most modern dairy operations. They allow separate groups of animals to be fed according to their nutrient requirements. Implementing TMR feeding normally increases feed intake, reduces digestive upsets, and improves the intake of low-palatability products. It is advisable to incorporate TMR feeding systems into any modernization plans. Housing that restricts animal grouping, and the high capital cost of TMR mixers, may preclude their use in some smaller operations.

When selecting a TMR mixer, the dairy manager may select mobile units that are used to deliver the feed to feed bunks, or stationary units that require conveying systems to deliver the feed mixture to the animals. Mobile mixers provide more flexibility, allowing feed ingredients to be loaded from different locations and animals to be fed at remote sites. Animal intake levels and feeding frequency determine the batch size for any animal group. The mixer should be sized so that the largest batch is 60–70 percent of the mixer's rated capacity. Mounting a mobile mixer on a truck is recommended if animals are fed at remote locations. There are many types of TMR mixers on the market (auger, tumble, reel, ribbon, etc.). Each type has advantages and disadvantages that the dairy manager must consider when making a buying decision.

Balancing Feed Storage Capacities with Expected Needs

Managers planning to modernize an existing operation should remember that direction of change is more important than speed. Often a phased approach involving several small changes can lead to the same end result, with less risk and trauma, than one large change. If the phased approach is selected, the decision maker should define the feed and feeding requirements at each phase and purchase only the required inputs as needed.

The Decision Process—An Example

The Smith family dairy farm currently owns 100 milk cows, 90 heifers, and 300 acres of land. Cows are housed in a tie-stall barn and milked with a pipeline milker, and manure is hauled daily. Their feeding program is based on baled hay and corn silage, which is stored overhead in the barn and in two tower silos. Two families operate the dairy, and together they have four children aged five to nine. One brother enjoys the dairy herd management chores, and the other enjoys the crop and heifer-raising aspects of the operation. One wife has record-keeping responsibilities, and the other has calf-raising responsibilities.

As these families review their operation, they realize they need to change, to keep up with the industry and to replace crop equipment and feed storage facilities that are no longer functional. At a family planning meeting, after reviewing their balance sheet and family goals, they agree that they are willing to borrow up to $1 million to make the improvements needed to become a long-term viable operation.

TABLE 10-1 Percentages of proposed forage capacity to be used for different herd sizes.

Storage Unit	Capacity ton DM	% of Capacity Needed 100 Head 500 ton	300 Head 1,500 ton	600 Head 3,000 ton	1,200 Head 6,000 ton
20′ x 50′ tower	130	26%	9%	4%	2%
120′ x 20′ x 10′ bunker silo	1,800	386%	129%	64%	32%
120′ x 20′ x 10′ bunker silo	1,800	746%	249%	124%	62%
120′ x 30′ x 12′ bunker silo	3,000	1,346%	449%	224%	112%

Working with a local consultant, they evaluate the feasibility of different herd sizes and facility types. They analyze the dairy, heifer, and crop enterprises separately to determine the feasibility of keeping each of the enterprises as they expand. They agree on a long-term plan to develop a 600-cow herd. This would be done in two phases, with the first phase having 300 head of milking cows. Since the heifer operation historically operates at a break-even level, and considering the shortage of cropland to support the expanded dairy, they agree to sell the operation's heifers and buy all replacement animals. Since their cropping equipment must be replaced, and they need additional feed, they agree to (1) sell existing cropping equipment, (2) purchase new haylage harvesting equipment, and (3) hire a custom operator to plant crops and harvest corn silage. They arrange to rent additional land from a neighbor for alfalfa and to buy corn silage in the field.

As they evaluate their existing feed storage, they realize that only one of their feed storage structures fits the new operation. This 20- × 50-foot tower silo is in good repair and is located where it can continue to be used. The forage storage analysis in Table 10-1 shows their plan for feed storage, based on the feeding program developed with their nutritionist. The table shows the different sizes of the forage storage units that would be added at each stage of the expansion. The percentage values shown are the combined amounts of the listed unit and those listed above it. An additional column was included to evaluate the effect of a later possible expansion to 1,200 head. The shaded values on the diagonal indicate that each additional storage unit provides sufficient storage to cover the needs of the dairy at that phase of expansion. It also shows that if all the units were built too early, there would be tremendous overcapacity that could not be economically justified.

Cropland Required

When determining the amount of feed needed to support a dairy herd, it is important to consider the number of animals and their size, ration ingredients, milk production levels, and all wastages expected. Crop amounts that account for expected intake, **harvest waste**, **fermentation waste,** and **feeding waste** must be estimated. Table 10-2 shows the amount of losses to expect with different methods of harvesting hay. High-producing cows are often fed 5–10 percent more TMR than their expected intake (Figure 10-10). To determine the forage needs for a 300-cow herd, the Smiths performed a cropland needs analysis (Table 10-3). Based on their expected feeding program, about 800 acres of cropland would be needed for the first phase of their expansion.

TABLE 10-2 Dry matter (DM) losses of hay expected with different harvesting techniques (Hoard's Dairyman, 1997).

	Harvest	Storage	Feeding	Total	Tons Needed/ Tons Fed
Haylage					
< 30% DM	2.0	21.2	11.0	34.2	1.45
30–40% DM	5.0	10.1	11.0	26.1	1.32
> 40% DM	11.5	8.2	11.0	30.7	1.38
Baled hay					
Rained on	32.6	4.0	5.2	41.8	1.63
No rain	17.4	3.6	5.5	26.2	1.32
Large bales					
Field cured	25.0	14.2	15.3	54.5	1.83
Acid cured	15.0	10.7	5.5	31.2	1.39

FIGURE 10-10 Animals should be fed with an expected 5–10 percent refusal rate, and uneaten feed should be removed daily.

TABLE 10-3 Acreage needed to supply feed for a 300-cow herd.

	Tons Needed as-fed	Tons Needed (DM)	Yield (Ton DM)/ acre	Acres Needed w/loss
Hay	902	767	4	192
Haylage	3,122	1,405	4	351
Corn silage	4,080	1,428	6	238
Total	8,104	3,600		781

TABLE 10-4 Total capital cost per ton of forage dry matter stored.

	Herd Size			
	75	150	300	600
Concrete stave	$192	$138	$132	$129
Bunkers	$152	$103	$88	$76
Piles	$63	$41	$28	$23
Bags	$88	$53	$35	$27

TABLE 10-5 Total annual cost per ton of forage dry matter stored.

	Herd Size			
	75	150	300	600
Concrete stave	$46	$36	$35	$34
Bunkers	$45	$37	$35	$32
Piles	$37	$32	$30	$29
Bags	$38	$32	$28	$27

Determining Forage Storage Type

To determine what forage storage system is best for a dairy, both the current and the long-term needs of the operation must be considered. If capital is limited, sometimes minimizing initial cost and incurring additional long-term labor costs can be justified. This is especially true if the initial purchase can be upgraded later (e.g., sides put on concrete pads to make bunkers). However, it is usually better to base the buying decision on the expected average annual total cost. Holmes (1996) showed the economic effect of different forage storage systems at different herd sizes. Table 10-4 shows the average capital cost, and Table 10-5 the average annual cost per ton of

TABLE 10-6 Expected initial costs of feed storage options.

	Construction Cost	Equipment Cost	Total Investment
Bags on stone pad	$27,280	$27,075	$54,355
Bunker with 8' sidewalls	$121,428	$11,025	$132,453

forage dry matter (DM) stored. The cost per ton decreases for all types as herd size increases, and piles and bags always require a substantially lower initial cost. When the annual costs are compared, this large difference disappears, and the decision may be based more on convenience or other related issues.

The Smiths did their research, and using all the values collected during the planning phase, they narrowed the choices to the two feed storage alternatives shown in Table 10-6. As they reviewed these options, they chose to use bags on a stone pad because they felt it gave them more flexibility and allowed them to keep some cash reserves in case they encountered cost overruns or additional start-up costs.

SUMMARY

When selecting a feeding system, as with other components of the dairy system, it is important to consider both the initial and ongoing operating costs. Each purchase should complement the needs of the dairy at the time of purchase and be usable during later expansion phases.

CHAPTER REVIEW

1. What are the three factors that determine a feeding program?
2. List the advantages and disadvantages of using silage bags for feed storage.
3. What are three benefits of using TMR?
4. List the three factors that account for wastage.
5. If 750 tons of baled hay are needed for feed, what is the total amount of baled hay that must be procured? Calculate tonnage amounts for both rained-on and dry baled hay.
6. The dairy profiled in this chapter opted to use bags on a stone pad to store forage, in addition to the existing 20- × 50-foot tower silo. What percentage of feed will the silo accommodate when the operation expands to 300 head? At that point, how many tons of feed will need to be stored in bags? At what point in the expansion do you think the dairy should upgrade its feed storage system?

REFERENCES

Hoard's Dairyman. (1997). Chore reduction for confinement free stall systems—A guide to improved returns for dairymen. *Chore reduction bulletin,* p. 2. Fort Atkinson, WI: Author.

Holmes, B. J. (1996). *Sizing and managing silage storage to maximize profitability.* [Department of Biological Systems Engineering document]. Madison: University of Wisconsin.

Holmes, B. J. (1997a). *Bagged silage or bunkers? Options for the expanding dairy farm.* [Department of Biological Systems Engineering document]. Madison: University of Wisconsin.

Holmes, B. J. (1997b). *Bagged silage or tower silos? Options for the non-expanding dairy farm.* [Department of Biological Systems Engineering document]. Madison: University of Wisconsin.

Kammel, D. (2000). Feeding facilities. In Bickert, W. G. et al., *Dairy freestall housing and equipment handbook* (7th ed., pp. 117–135; *MWPS-*7). Ames, IA: Midwest Plan Service.

Chapter 11

Manure Handling Options

KEY TERMS

3-basin storage system
automatic alley scraper
composting
diffuser
manual scraping
manure separator
sand separator
slatted (slotted) floor
solids separator
turn wheel
water flush

Objectives

After completing the study of this chapter, you should be able to

- identify the facility requirements of manual scraping, automatic alley scraping, water flush, and slatted floor manure removal systems.
- identify the size and soil requirements of various manure storage facilities.
- understand the relationship between manure handling systems and crop size, soil health, weather conditions, and labor efficiency.
- understand how bedding materials can impact all components of a manure handling system.
- accurately project the type and size of manure handling system required by an existing dairy's modernization plan.

The selection of a manure handling system for a dairy should be based on the expected long-term size of the operation, the amount and location of available land, and the types of crops to be grown. Local weather and land characteristics can influence the frequency with which manure storage units can be emptied, thereby also influencing storage size requirements. The system must be environment- and neighbor-friendly, labor efficient, and have an initial cost that can be supported by the expected long-term milk price. It must include methods to collect the manure, move it to storage, process it if needed, keep it until it can be disposed of, and then remove it and dispose of it. The size and type of structures and equipment selected must be compatible with the type of bedding materials used and the weather conditions that can be expected.

Freestall Barn Manure Collection and Removal Systems

Producers normally have the following five manure collection and removal systems to select from.

- **Manual scraping** with a tractor or skid steer
- **Automatic alley scraper**
- **Water flush**
- **Slatted (slotted) floor** over gravity channel
- Slatted floor over storage tank

Manual scraping with a tractor or skid steer is often the choice of owners of smaller dairies, and those who are trying to minimize the initial capital cost associated with expansion. Almost 80 percent of the respondents to a recent survey of Wisconsin producers (Palmer & Bewley, 2000; Table 11-1) who increased herd size by at least 40 percent reported using this method of

TABLE 11-1 Average satisfaction score and cost of manure removal systems as reported in the 1999 Wisconsin Dairy Modernization Survey (Palmer & Bewley 2000).

	Tractor Scrape	Alley Scrapers	Slats	Flush
Number of herds	189	26	17	5
1998 median herd size	205	283	370	545
Cows per FTE	42[b]	50[a]	43[ab]	38[ab]
Manure management*	3.55[b]	4.39[a]	4.65[a]	5.00[a]
Bedding usage and cost*	3.95[b]	4.39[a]	4.41[a]	4.20[ab]
Cost per stall	$986[b]	$1,111[b]	$1,458[a]	$1,095[ab]

[a,b,c,d]Means within rows with different superscripts differ ($P < 0.05$).

*Average satisfaction reported on a scale from 1 (very dissatisfied) to 5 (very satisfied).

TABLE 11-2 Total outlay and annual cost for manure handling and storage systems for freestall barns with no insulation.

Manure Handling System	Manure Storage*	Initial Cost ($/cow)	$7/hr Labor Annual Cost ($/cow/yr)	$14/hr Labor Annual Cost ($/cow/yr)
Tractor scraper	CL basin	382	158	176
Alley scraper	CL basin	444	160	164
Water flush	3 CL basins	737	216	220
Slatted—gravity channel	CL basin	845	207	208
Slatted—over tank	C tank	1,033	215	216

*CL—Concrete Lined, C—Concrete

manure removal. These producers reported spending less per cow for housing but were not as happy with manure management as producers who chose other, more expensive options. Many of these producers also chose sand bedding, which is probably why they also were less happy with bedding usage and cost. Producers who chose other manure handling systems avoided sand bedding because of its tendency to settle out in storage, increase equipment wear, and increase manure removal costs.

Table 11-2 shows calculated expected initial and annual costs of the different manure handling systems (Holmes, 1995). These estimates were calculated assuming uninsulated barns, seven months' manure storage (except slatted—over tank, which was one year), and the equipment and labor required for normal and winter manure removal. These values show that manual scraping with a tractor scraper had the lowest initial investment per cow, but when the yearly labor cost was considered, automatic alley scrapers were very competitive. One of the costs associated with tractor scraping that was not included was the damage that can be done by hired employees to buildings, gates, equipment, and animals with tractor scrapers. The other systems had higher initial and annual costs but should be considered nonetheless, based on the convenience they offer (as shown by the user satisfaction scores from the 1999 Wisconsin Dairy Modernization Survey).

Automatic Alley Scraper Systems

Automatic alley scrapers have a drive unit that pulls a scraper blade up and down each manure alley, dropping the manure at the center or end of a freestall barn (Figure 11-1). This drop should be protected with gates, grates, or slotted floors to prevent animals and people from falling into the manure collection channel. If grates or slotted floors are used, they must be designed to support sticky or partially frozen manure associated with colder weather. Often alley scrapers are set to remove manure 8–12 times per day.

FIGURE 11-1 Automatic alley scrapers offer an economical method of removing manure from freestall barns. Manure drops should be enclosed to protect animals and people.

FIGURE 11-2 In cold climates, reception pits at the end of freestall barns must be well insulated to prevent them from freezing.

The longer the period of time between scraping, and the longer the barn length, the greater the quantity of manure the scraper will pull. If this pile of manure is too deep, it may overflow into the stall beds or cover the lower portion of a cow's legs when she is standing in the alley. Multiple scraper blades can be included for long barns, which allow manure to be shuttled from one blade to the next. In cold weather, the alley scrapers may be run more frequently to minimize manure build-up. If a reception pit will be constructed at the end of a barn, it should be insulated to minimize freezing (Figure 11-2). If extremely cold weather occurs frequently, the manure handling system should be designed so that frozen manure can be removed from the barn. Many producers remove the alley scraper cables and use a tractor scraper during these times. Others have installed in-floor heating

FIGURE 11-3 Some producers have successfully installed in-floor heating to prevent manure from freezing.

systems that prevent the floor from freezing (Figure 11-3). Other producers have closed part of the barn's ridge or eve openings to minimize heat loss and to increase the barn's internal temperature. This is not recommended because it may be unhealthy for the animals or damage the barn. Automated alley scrapers require routine maintenance (greasing and cable tightening) to operate properly. In cold climates, when designing a new freestall barn that will initially or eventually incorporate automated alley scrapers, reserve space inside the barn for the drive unit and **turn wheels** (Figure 11-4). This will allow operators to maintain them easily during bad weather.

Water-Flush Systems

Water-flush systems normally use large tanks located outside and at the end of the barn, both to store a large volume of water and to develop enough pressure for the water to remove the manure from the manure alleys. The flush water is delivered by pipes to the ends of the alleys with enough slope (a minimum of 1 percent) to remove the manure. Water is released through **diffusers,** flows down the alley as a wave picking up manure, and is caught in a small reception pit at the center or end of the barn. (Some flush systems do

FIGURE 11-4 If automatic alley scrapers are to be used in cold climates, space should be provided at the end of the freestall barn for the drive units and turn wheels.

not use tanks but pump water directly from a lagoon. These systems require less barn floor slope but take more time and water to remove the manure.) From this reception pit, the flushed manure is taken to the manure storage area (where some of the solids can be separated using a **manure separator**) or placed directly into a manure basin. The high initial cost associated with the water-flush system that is shown in Figure 11-2 is largely owing to the selection of a **3-basin storage system.** With a 3-basin system, the first basin is used for settling out most of the manure solids; the liquid part flows to the second basin, which allows additional settling; and the liquid that flows to the third basin is used to fill the flush tanks or is pumped directly. Water-flush systems are common for dairies with large freestall barns in warm climates and have also been used successfully in cold climates. In cold climates, a secondary manure handling system must be used during extremely cold conditions. Most flush system designs require more storage capacity because of the addition of the water needed to drive the system. The additional water used by these systems makes hauling manure more expensive, so farm owners with land located far from the dairy may want to consider an alternative system.

Slatted Floor Systems

With slatted floor systems, manure is forced through slatted floors in the manure alleys by hoof traffic and gravity. With a gravity-channel system, manure is temporarily stored under the slatted floor, but gradually flows over one or more step-dams to a reception pit area and then is channeled to an outside storage basin (Figure 11-5). Since manure is constantly moving out of the barn, there is minimal concern with manure gas build-up. Cows are often quite clean with this system because most of the manure and urine is quickly removed from the barn floor surface.

FIGURE 11-5 Slatted floor over gravity-channel manure handling systems allow manure to drop through floor openings into a channel below the floor and then flow out of the barn to permanent storage. They can minimize labor, minimize concerns with manure gases, and provide an environment conducive to clean cows, but may have a higher initial cost than other systems.

In contrast, under-the-barn manure storage systems usually have capacity for a full year's manure generated by animals in the barn. Producers who do not have space on the site for an open basin or tank often select such systems. This type of system usually has a higher initial cost than other manure handling systems, but it has the lowest labor requirement. Family farms that want to milk more cows without hiring labor will sometimes choose this type of system, even though the average annual cost is higher. In that case, the producer must ensure that the barn is properly ventilated to minimize the chance of manure gas buildup, which can depress animal intake and may be detrimental to animal and human health. With both slatted floor systems, alleys are normally narrower to increase cow traffic and the amount of manure forced through the slots in the floor.

Manure Transport Methods

With all but one of the manure collection systems mentioned, manure must be moved from some location in the center or at the end of the freestall barn to long-term storage. Manure is normally collected in a reception pit and is then pumped or flows by gravity to a storage unit. Selecting a site that allows the freestall barn to be built high enough above the manure storage pit to allow manure to flow by gravity is highly recommended because it saves the initial, operating, and maintenance costs of pumping equipment.

The reception pit design should be based on the type of manure and bedding materials to be used. Often producers will build a concrete container about 10 feet wide and 10 feet deep across the width of the barn (Figure 11-6). This has the advantage of being able to hold several days' manure and can be used for temporary storage if modified daily haul is desired. This design is not recommended if sand bedding will be used,

FIGURE 11-6 Manure reception pits in the center of the barn can be used to collect manure, provide short-term storage, and feed transfer systems that take the manure to long-term storage.

FIGURE 11-7 If sand bedding will be used, the center reception pit should be narrow and easily accessible for cleaning.

because the sand in the manure will settle to the bottom and quickly fill the pit. If sand bedding is to be used, a narrow channel (18–24 inches wide) will minimize sand settling, but provisions should still be made to allow sand buildup to be removed if the channel becomes clogged (Figure 11-7). These concrete containers are expensive to build because of the amount of

FIGURE 11-8 Manure flumes, made from plastic culvert pipe, make an inexpensive manure collection and transport system across freestall barns.

reinforcement rod needed. A less expensive way to move manure across the barn is with a round flume buried under the floor (Figure 11-8). These flumes are often made using 24-inch plastic sewer pipe, with openings at the end of each manure alley that allow the manure from that alley to drop into the pipe and flow away. These work well for freestall barn manure, unless it is sand-laden.

Manure Storage Options

The most common long-term manure storage options include the following:

- clay-lined storage basins (lagoons)
- concrete- or flexible-membrane–lined storage basins
- walled enclosures (vertical-sided concrete tanks, aboveground storage tanks, or under-the-barn storage tanks)

Early in the expansion planning process, after the long-term herd size goals have been defined, soil borings should be done to determine the soil composition and water table heights of potential sites. It is also advisable during this time to have any potential site surveyed and a topographical map generated. Then the proper placement of the manure storage and building complex can be determined. The topographical map shows the height at each point on the site and can be used to determine the amount of surface material that must be moved to make a properly leveled and sloped surface to accommodate the facility. The soil borings will provide information on the amounts and types of materials under the surface—amount and quality of clay present, permeability of the soil, presence of water, and presence of rock. Each of these has an impact on the proper location of the manure pit and facilities, and the cost of site preparation. If soil type and water level conditions are satisfactory, a site where a clay-lined storage basin can be constructed will provide the cheapest manure storage. Clay-lined storage basins have sloped interior side walls (often 2:1 or 2.5:1) covered with clay, which prevents seepage of manure and contamination of ground water. If a clay-lined storage basin is not possible, the next cheapest option normally is a storage basin of the same design, but sealed with concrete or synthetic flexible-membrane materials. Local zoning, a high water table, or other factors may require a vertical-sided walled enclosure, which is often more expensive to build because of the extra reinforcement rod needed. Whatever storage design is selected, the implications of agitating and removing the manure, and the basin's location relative to local housing (for odor control) should be considered.

Sizing Manure Storage

The basic things to know when sizing manure storage are the number of animals, the bedding materials to be used, and the length of time storage is needed. Another factor that influences storage requirements is the amount of water from parlor milkline washing, parlor and holding-area flushing, groundwater runoff, and precipitation falling on the storage. Local, state, and federal regulations that apply to your location should be researched to ensure that your planned system will be in compliance.

Many dairies, when they expand, build their manure storage systems too small. This can cause several major problems; for example, it increases the frequency and cost of emptying the storage, and it may necessitate manure removal when cropland is not available for manure application. Experience would indicate that a minimum of 12,000 gallons of storage per cow per year is needed for large modern dairies. The size of the animal and its production level greatly affects the amount of manure generated. High-producing dairy cows eat more and produce more manure. Sand bedding has two effects: first, it adds a large volume of material; second, because it is difficult to agitate and remove, it may actually decrease the holding capacity of the storage over time. Facilities should be designed and sites landscaped so that clean water is channeled away from manure storage. Manure storage should be designed to allow easy access to manure-agitation and -removal equipment. Figure 11-9 shows agitation equipment used where concrete access pads are not provided near the pit. This type of equipment is uncommon and expensive. The pit should be narrow enough to allow equipment to properly agitate the manure, to break up crusts and put residues back

FIGURE 11-9 A manure storage pit, built without concrete access pads, being agitated by long-reaching equipment.

FIGURE 11-10 If a dairy uses sand bedding and a single manure storage unit, the pit should be designed to allow access for removal of settled sand.

into suspension. For sand-laden manure, a pit should have a concrete bottom and access ramp to allow equipment to enter the pit and remove sand deposits (Figure 11-10). Safety is a key factor when considering manure storage systems. Fencing open storage and providing proper ventilation for under-building storage are musts. Figure 11-11 shows manure storage that has concrete agitation pads and is protected by fencing.

FIGURE 11-11 Manure storage units should be fenced to protect people and animals.

Other Considerations

Thinking through the labor and cost relationships of manure handling require that you consider not only removal and storage but also disposal of manure. Most large producers now hire custom manure haulers to empty their manure storage units. These professionals normally charge a set-up fee and a cost per unit removed. This unit cost can vary based on the material being handled. For example, fees are normally higher for removing sand-laden manure because of the extra costs associated with it. Extra storage capacity can minimize the frequency with which storage must be emptied and result in lower set-up costs. The extra capacity also provides flexibility when working around the schedules of custom contractors.

Location of cropland, and field slopes, can influence the choice of manure hauling equipment and the associated cost. Manure spreaders work well for small operations, and may be the logical choice for removing bedding-pack materials. Manure tankers can provide uniform manure application but may contribute to soil compaction. Flexible-drag-hose systems support fast manure removal with minimum soil compaction but work best when transport distances are less than two miles. Odor during and after application can be a real concern to neighbors and can be minimized by

FIGURE 11-12 This screen separator is one of many different types of manure separators on the market.

FIGURE 11-13 Producers who use separated and composted manure solids for freestall bedding are pleased with their results.

incorporating manure into the soil. The amount of land needed to dispose of manure continues to be a major concern of producers with large herds, and will increase in significance as nutrient management systems shift from a nitrogen-limiting system to a phosphorous-based system. This has the long-term implications of more land needed per animal unit and longer manure transport distances. Manure modification, using manure separators (Figure 11-12), **composting** (Figure 11-13), methane generation equipment (Figures 11-14 and 11-15), and so on, is becoming more popular, and interest in such systems will continue to grow as large herds face the growing demands for control of odor, surface and groundwater contamination, and road littering and the impact of new nutrient management rules. Two types

FIGURE 11-14 Methane generators use anaerobic digestion to process manure, producing methane gas. The gas is used to power generators, producing electricity that can be used by the dairy or sold back to power companies. Digestion takes placed in concrete pits that are covered to contain the gases produced.

FIGURE 11-15 Generators take the methane gas generated by methane digesters and produce heat and electricity.

of manure separators are currently in use. **Solids separators** are designed to remove the solid portion of manure, and **sand separators** are used to separate sand from manure. Separating the solid and liquid portions of manure can contribute to nutrient control on the farm by allowing the solid portion to be transported longer distances and the liquid portion irrigated on nearby land. Separated sand is often reused, thereby decreasing the cost of bedding. Additional benefits of manure modification are numerous and often can help justify their cost. An example is the use of manure solids for freestall bedding, which many producers employ successfully.

Methane Digestion

Methane digesters use anaerobic digestion to produce energy in the form of methane gas (CH_3) and heat. For example, the manure from each 100 cows can produce up to 480 Kwh per day of electricity, worth over $10,000 per year at $0.06 per Kwh. The heat generated can be used to heat water, and so on, for additional on-farm savings. A reduction in odor is also achieved, and this is often a major reason for installation of these systems. Research at Iowa State University showed over 80 percent reductions in hydrogen sulfide and odor in digested manure when compared to undigested manure. Manure volume and nutrient content is essentially unchanged. Overall nitrogen content is reduced slightly, but phosphorus content is not changed. Such digesters are expensive ($600–1,000 per cow) and currently do not become economically feasible until electricity prices exceed $0.10 per Kwh. The biggest obstacle to overcome with digesters is the management required. They are very complex electrical, mechanical, or hydraulic devices. Covered, insulated tanks with good heat exchangers are required in northern climates. Some sacrifices may also be needed; for example, plug-flow systems, typically used on dairies, would not work with flush systems, and wood chips and sand bedding can not be run through digesters.

Vacuum Loading

Vacuum loading is a relatively new method of handling manure. Initial reactions from producers using this technology are very positive. Manure is vacuumed from manure alleys (Figure 11-16) and loaded directly into truck-mounted or trailer-mounted tractor-pulled units (Figure 11-17). The manure can then be hauled to local or remote lagoons for storage, added to compost piles, or applied to fields. Different models of this equipment are currently marketed that have 3,750–12,000-gallon capacities. The manufacturers' promotional materials state that these units can load up to 4,000 gallons per minute. They can load manure containing sand bedding as well as most other commonly used bedding materials. The design of facilities for producers considering this manure handling option is simplified because there is no need for reception pits, cross channels, and manure storage near animal housing. No cost–benefit analysis is currently available for this new technology.

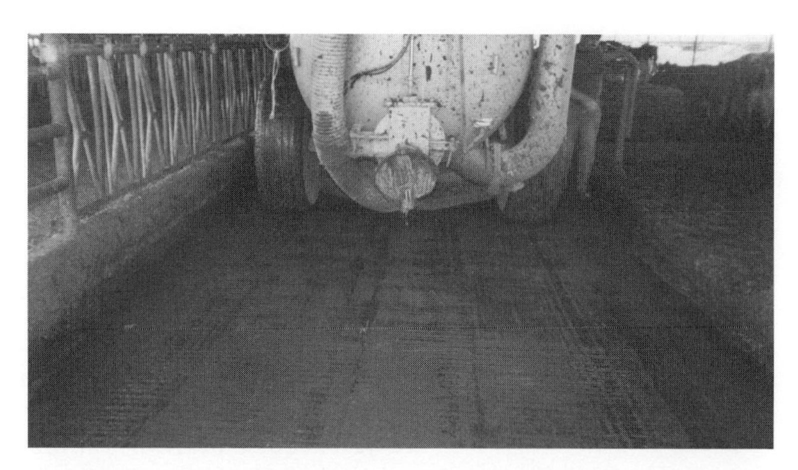

FIGURE 11-16 Manure alleys in freestall barn after removal of manure by vacuum-loading unit.

FIGURE 11-17 Trailer-mounted vacuum-loading unit used to remove manure from freestall barn alleys.

SUMMARY

Choosing the proper manure handling for an expanding dairy is very complicated and is often a compromise between what can be justified financially in the present and what the long-term needs and economics of the dairy require. Making the proper decision is complicated by the unknown future influence of government and neighbors. Manure is a resource that, if properly used, can decrease feed production costs. Part of expansion planning should include an evaluation of the value of manure for the geographic location being considered. No matter where the dairy is located, the owner must ensure that it is environment- and neighbor-friendly and labor-efficient, and see to it that any proposed changes do not negatively impact the long-term viability of the operation.

CHAPTER REVIEW

1. List three important steps in manure handling.
2. List the advantages and disadvantages of the following manure collection and removal systems: manual scrapers, automatic alley scrapers, water-flush, and slatted-floor systems.
3. Diagram the manure-collection and -removal systems listed above. Be sure to include key features such as manure alleys, freestall curbs, reception pits, and storage sites.
4. What is the minimum number of gallons of manure per cow per year that a manure storage facility at a large dairy should accommodate?
5. List the advantages and disadvantages of clay-lined and flexible-membrane–lined storage basins.
6. Explain how excess sand affects a manure handling system. Be sure to identify where sand originates and how it is disposed of.
7. Why do farmers use custom manure hauling, given its expense?

REFERENCES

ASAE Agricultural Sanitation and Waste Management Committee. (1997). Control of manure odors. [Standards document EP379.2]. St. Joseph, MI: American Society of Agricultural Engineers.

Bewley, J., Palmer, R. W., & Jackson-Smith, D. B. (2001). A comparison of free-stall barns used by modernized Wisconsin dairies. *Journal of Dairy Science, 84,* 528–541.

Holmes, B. J. (1995). *Manure handling costs—Freestall barns to field.* [Department of Biological Systems Engineering document]. Madison: University of Wisconsin.

Holmes, B. J. (2000). *Dairy manure storage planning to meet current and future needs.* [Department of Biological Systems Engineering document]. Madison: University of Wisconsin.

Lorimor, J. (2004, January 10). Is an anaerobic digester right for you? *Hoard's Dairyman,* 22–23.

Palmer, R. W., & Bewley, J. (2000). The 1999 Wisconsin dairy modernization project—Final results report. Madison: University of Wisconsin.

University of Wisconsin Cooperative Extension. (1998). *Manure management choices for Wisconsin dairy and beef cattle operations.* [GWQ 024]. Madison: Author.

Chapter 12

Animal Acquisition

KEY TERMS

AI progeny-tested bull
AI young sire
biosecurity
body capacity
BVD (bovine virus diarrhea)
chronic staph
contagious organism
dystocia
hairy heel wart
inbreeding
Johne's disease
jumper bull
mastitis
net merit
PTA fat
PTA milk
PTA protein
quarantine
salmonella
springing heifer
transition difficulty
uddering up

Objectives

After completing the study of this chapter, you should be able to

- list the characteristics of a healthy and an unhealthy heifer.
- list the diseases that commonly afflict herds.
- identify which animal selection methods can accurately predict the quality of an animal and which cannot.
- identify transition problems that can be encountered when bringing animals into an existing or expanding operation.

Many producers who want to modernize their dairy operations would like to maintain a closed herd and grow from within, but this is not feasible because of the large increases in herd size needed to cost-justify the investment in new facilities. Cash-flow analysis of dairies shows that the quicker a dairy can be filled, without introducing animal or labor problems, the better the financial health of the dairy will be. This rapid growth in herd size necessitates the purchase of animals from outside sources. The selection and handling after purchase of these animals can greatly affect the level of health problems and culling rates in expanding dairies. Developing an animal acquisition plan early in the planning process can help avoid potential disasters.

Factors to Consider

To maintain or develop a high-producing, highly profitable herd during expansion many factors must be considered. Evaluate the possible sources of animals, select the right type of animal, and then work out the correct method to handle animals as they are introduced into the herd. Healthy, high-genetic-potential, well-cared-for animals are a must for a highly profitable herd.

Animal Sources

Cattle are normally purchased privately from other producers or bought though common sources such as auctions, sale barns, or cattle dealers. Almost half of the respondents to the 1999 Wisconsin Dairy Modernization Survey indicated they grew from within (Bewley, Palmer, & Jackson-Smith, 2001). Of those who bought animals, 44 percent bought privately from other producers, 22 percent purchased at auctions, 6 percent at sale barns, and 27 percent from cattle dealers. Table 12-1 shows the types of animals they reported buying. An almost equal number of respondents indicated

TABLE 12-1 Types of animals purchased by expanding dairies.

	Number of Herds	%
Bought bred heifers before they calved	198	66
Bought mature animals	188	63
Grew from within	145	48
Bought bred heifers that had recently calved	64	21
Bought calves or heifers and raised them	50	17

that they bought heifers before they calved and mature animals. Buying first-lactation animals after they calve is becoming more common; 21 percent of the respondents used this practice, which offers several advantages to start-up dairies (or dairies that need to increase herd size drastically). Since these animals have calved and have been milked several days or weeks, they are accustomed to being milked, their milk production level is known, and the quality of their udders can be determined. This minimizes the labor requirement of the buyer, decreases the culling rate, and eliminates the delay between animal purchase and milk flow to the tank. Since the seller supplies these extra services to the buyer, the price is normally $100–200 higher to compensate for the additional labor and risk associated with calving problems or the decreased value of blemished animals. If you plan to purchase animals from dealers or agents, it is wise to investigate their reputation. Can and will they deliver as promised? Develop a list of suppliers. Discuss your needs with other dairies, and ask about their experiences with suppliers. Call suppliers to determine if they can meet your needs. Interview suppliers or visit their sites to increase your comfort level. If a large number of animals will be needed, selecting two or three suppliers will decrease the risk of non-delivery. Increasing the number of sources of animals can increase health risks and, consequently, the importance of precautionary **biosecurity** measures. If complete herds are to be sourced, a commission rate of about 5 percent is normal, whereas a fee of about $25 to select a heifer can be expected. This fee will increase if you place restrictions on the type of animal desired or place the supplier at risk in the transaction. For example, if animals must be tested for **BVD (bovine virus diarrhea)** or **Johne's disease,** an additional $50–100 may be added to cover testing costs and the liability for rejected animals. Requiring the animals to have sire or dam identification can limit the sources of heifers and increase their cost.

Animal Types

Age, Size, and Condition

The decision of what age animals to buy often is a function of the amount of time before they are needed, and the owner's situation. If you have the feed, labor, financial resources, and facilities to raise the animals, buying them early gives you flexibility to buy when the market is right, raise the heifer according to your specifications, and have animals adjusted to the dairy's housing and feeding program. It does offer a challenge, however, in that animals will be ready to calve at a specific time, which sets a hard deadline on when the facility must be ready.

Buying animals a few weeks or months before calving allows them to be on site, and allows management the ability to quarantine, treat, and acclimate them before the traumatic calving time. A minimum of 60 days should be allowed for acclimatization of a **springing heifer** and should be reflected in the price paid for the animal. Remember that these animals eat a lot of feed as they approach calving time.

Whatever the age of the animals purchased, they should weigh or be on track to weigh 1,350–1,450 pounds before calving. In a rearing situation, age at first calving is an important factor in reducing rearing cost, but in this situation age has less importance. In fact, there may be a slight benefit of heifers that are a little older, as long as the age is within a reasonable range (i.e., expected calving age 23–28 months). Animals that will mature into average-size cows are preferred, since they will fit the parlor better and have been shown to be more efficient than extremely large animals. Weight for age is a principle to follow when selecting younger animals; it refers to the animal's skeletal size (i.e., its height and body length) in relation to its age. Such guidelines will help you select animals that will mature to the size desired. To take a 90-lb heifer calf to 1,350 lbs in 24 months requires an average gain of 1.8 lbs per day.

Body condition is of paramount concern; fat heifers (body condition scores greater than 3.5–3.75) should be avoided because they may have **transition difficulties** (metabolic problems such as **dystocia,** etc.), and the excess weight indicates that the grower who raised them did not manage them properly and suggests that other problems may appear. The price of animals should reflect their body condition and be adjusted for any additional costs associated with it.

If heifers are expected to calve shortly after purchase, a premium can be paid to have known AI breeding dates. Some experienced buyers suggest that $50–75 is reasonable. One major problem encountered by expanding dairies is buying heifers that do not calve when expected. Extra time on site increases feed cost and has the potential to result in overconditioned heifers, which can lead to increased problems during and after calving. Rectal palpation of animals can determine pregnancy status but is not an accurate method of establishing expected calving dates. To select only animals that are **uddering up** helps minimize the problem of unknown breeding dates but is not much help when buying large numbers of animals and can result in animals being on site insufficient time to adjust properly.

Body Conformation

Since heifers do not have developed udders, little importance can be placed on this trait unless some obvious flaws exist. Inspect heifer calves and yearlings to ensure that there is no overdevelopment or fatty udders. Avoid animals with more than four teats, and watch for enlarged quarters, which may indicate that the animal was suckled. Older animals should be inspected for undesirable teat size and placement, side-leaking teats, light quarters, and unbalanced udders.

Feet and legs are extremely important since they relate to an animal's longevity and mobility. The front and rear legs should be set relatively wide apart. Rear legs should have some angle to the hocks when viewed from the side. Animals with pasterns that are too weak or too straight seem to have more lameness. Long claws, **hairy heel warts,** and feet covered with caked manure and dirt indicate lack of attention by the grower. **Body capacity** refers to the animal's body length, depth, and width; width of chest; spring of ribs; and fullness of crops. Body capacity indicates the ability of the animal to consume large amounts of feed to support rapid and efficient growth and production. Frail animals should be avoided since they may not have the vigor and stamina to compete in modern dairy facilities.

Genetics

Everyone knows that the genetic potential of an animal limits its ability to produce, reproduce, and stay in the herd (Figures 12-1 and 12-2), but what should you look for? How much should you pay when you need to select a large number of animals at one time to fill a new or expanded facility? In recent years, many new dairies have selected heifers from large heifer ranches, or cattle buyers who consolidated them. The results have been mixed with respect to production levels and culling rates.

Animals with unidentified parentage are likely to be of lower quality and have higher cull rates than sire-identified animals. Commercial channels often have a high proportion of heifers sired by **jumper bulls** of unknown genetic merit, or that come from producers who sell their poorer animals.

The U.S. Department of Agriculture performed within-herd comparisons of milk, fat, and protein yields of grade cows with and without sire and dam identification (USDA, 2001). Animals with known or unknown dams were not significantly different, but the effect of an unknown sire was significant. Cows with identified sires produced, on the average, 370 lbs more milk, 11 lbs more fat, and 10 lbs more protein. Using statistical

FIGURE 12-1 Selecting medium-sized cows with good feet, legs, and udders and the genetics for high production, supports the long-term profitability of an expanding dairy.

FIGURE 12-2 Cows that lack the strength to compete in modern facilities may have to be culled early, decreasing the profitability of a modern dairy.

models, they estimated that these animals would have 1.8 additional months of productive herd life and 0.18 lower somatic cell scores. Using 1996 economic conditions, they estimated that an identified-grade cow would be $166 more profitable than a non–identified grade cow. Being registered added another $75 to the animal's productive value, giving registered animals an expected $242 more lifetime profit than non–sire-identified animals. It should be noted that it is not the identification itself that is important; proper identification indicates that other good management practices, which help develop a superior heifer, have been followed. The average value of AI-sired animals is considerably higher than non–AI sired animals. Table 12-2 shows the USDA Animal Improvement Program's latest evaluation results for **PTA** (predicted transmitting ability) for **milk, fat,** and **protein:** the average **net merit** of **AI progeny-tested** bulls is + $304, of **AI young sires** is + $190, and of non-AI sires is + $64, where net lifetime profit relates to the genetic base year average cow (the mean genetic value for cows of a specific breed in a given year). So the expected difference in net lifetime profit between daughters of the proven AI bulls versus non-AI bulls would be $240, and the difference between the AI young sires and non-AI bulls would be $126. If you buy heifers that are sired by the AI bulls, then you would expect $126–240 higher lifetime profit than from heifers with unknown parentage. Some nonidentified animals might be the result of AI, but it is doubtful, because a breeder who uses AI will probably make the effort to record it. The genetic advantage of AI sires has the same economic effect on the unborn calf's value.

A second advantage of AI breeding relates to **inbreeding.** We know that a 1 percent increase in inbreeding costs $23 in lifetime net profit (Smith & Cassell, 1998). We also know that well-conceived computerized mating programs can reduce inbreeding in replacement heifers by about 2 percent in the next generation. Buying animals with no ancestry information makes

TABLE 12-2 U.S. Holstein bull production evaluation results (USDA, 2001).

Active AI bulls	
PTA milk	+ 1,112
PTA fat	+ 36
PTA protein	+ 35
Net merit	+ $304
AI bulls born in the last 8 years (including those that were culled after their progeny test)	
PTA milk	+ 667
PTA fat	+ 24
PTA protein	+ 24
Net merit	+ $190
Non-AI bulls born in the last 8 years	
PTA milk	+ 114
PTA fat	+ 7
PTA protein	+ 6
Net merit	+ $64

it impossible to control inbreeding, whereas loss of milk production due to inbreeding can be minimized by buying replacements that are AI sired.

Buying animals bred to AI sires helps determine the genetic merit of the animal's offspring and minimizes heifer calving problems, since calving ease information is available with AI sires.

Health History

Buying animals from known sources and implementing sound biosecurity practices are crucial for rapidly expanding herds. Biosecurity practices help prevent the introduction of disease and pathogens to the operation, and control spread within the operation. Any time an animal from an outside source enters the herd, it brings a risk that **contagious organisms** will be introduced and affect the existing herd. The more sources of animals, the greater the risk.

If mature animals are purchased, cows with **chronic staph** or a history of serious **mastitis** should be avoided. Working with your veterinarian, you should develop a health screening protocol to follow. This protocol should be decision-focused and flexible enough to adapt to the unique situation of the individual dairy. For dairies whose primary income is from milk shipped, diseases resulting in loss of production or premature culling take high priority. For U.S. herds, contagious mastitis, Johne's disease, BVD, and **salmonella** species are major concerns. (Operations that sell semen or embryos are likely to also have concerns about BLV and blue tongue because of international trade restrictions.) If mature animals are bought from existing

TABLE 12-3 Biosecurity practices reported by expanding dairies.

	Number of Herds	%
Visually inspected animals before purchase	238	91
Increased level of vaccination in existing herd	177	67
Vaccinated incoming cattle after moving them	134	51
Vaccinated incoming cattle before moving them	129	49
Examined individual somatic cell count records	110	42
Isolated animals after moving them	72	27
Examined individual cow health records	67	26
Blood-tested animals before purchase	56	21
Did bulk tank cultures before purchase	39	15

herds, bulk tank cultures should be taken, health records checked, and the herd's veterinarian questioned about their health history. Individual animal testing (for diseases for which the test has a high sensitivity level, such as BVD) or herd testing (for diseases for which the test has a low sensitivity, such as Johne's) is recommended.

Table 12-3 shows the biosecurity practices reported by respondents to the 1999 Wisconsin Modernization Survey (Bewley et al., 2001). It appears that most producers visually inspected the animals (91 percent), vaccinated their existing herd (67 percent), and vaccinated incoming cattle (51 percent after and 49 percent before movement), while fewer producers isolated (27 percent) and blood-tested (21 percent) incoming animals.

The health protocol should define not only the practices for selecting animals but what should be done before and after new animals arrive at the dairy. Animals entering the dairy should be put in **quarantine** to reduce the risk of some diseases (such as BVD), but isolation is not effective for control of Johne's disease. Vaccinating the existing herd and new animals before they enter the herd can reduce the risk of clinical disease.

Transition Problems

Early in the expansion planning process, sources of animals should be investigated. It is critical to know where the animals will come from and when they will begin to freshen. Coordinating when housing and milking facilities will be ready for occupancy with expected animal calving dates is a must. Empty facilities are unprofitable, and having heifers ready to freshen without a place to be housed or milked is one of the most severe start-up challenges. Waiting until a facility is almost ready can be dangerous, in that the supply or cost of animals may be unpredictable. Contracting for animals early in the planning phase helps ensure that the animals will be there, and the price can often be fixed. To buy heifers early at an established price normally requires a down payment of about 10 percent of the agreed-upon price.

Many producers have had transition problems moving mature animals into a new facility. Cows taken from tie-stall to freestall barns often have high culling rates. This is probably caused by the added stress placed on the older animal learning to compete in a new environment, and may be exacerbated by the amount of time spent on concrete floors. Close observation of these animals, and relocation to bedded pack for a period of time if problems occur, can help reduce the culling associated with the transition. Training older stall-barn cows to enter a milking parlor can be a challenge. Expect problems when cows that have been trained in a herringbone parlor must be taught to use a parallel parlor. Many producers who have made the transition to modern facilities recommend purchasing heifers because they are easier to train, have a longer life expectancy in the herd, and may be less prone to bringing diseases to the dairy. Mature animals or heifers that have freshened before arriving at the dairy allow the buyer to inspect the animal's udder and production level, however.

Cows and heifers may have problems eating through self-locking manger stalls or lying in freestalls. Selecting animals that have experience with freestalls or self-locking manger stalls, and training animals as soon as possible before they freshen, will help avoid or minimize these problems.

SUMMARY

Obtaining a sufficient number of the right type of animal, at the right time, to increase herd size during expansion is a very important part of the planning process. Well-grown healthy heifers with good genetic backgrounds can greatly increase the long-term profitability of the dairy. Researching suppliers, developing proper biosecurity protocols, and designing comfortable, labor-efficient animal facilities also play a major role. Do not spend $1 million on a new facility and try to pay for it with $800 springing heifers.

CHAPTER REVIEW

1. How much should a cow weigh when calving for the first time? What is the ideal age of a first-time calving heifer?
2. List three benefits of buying first-lactation animals after they calve.
3. List five physical traits that are useful in evaluating a cow's health. Write a brief explanation of what each trait reveals about the animal.
4. List three diseases for which biosecurity measures should screen when cattle are brought into a dairy.
5. Identify three transition problems that might be encountered when bringing an animal into an existing or expanding operation.
6. What are the benefits of buying animals from reputable agents, fellow producers, or the AI?
7. If 27 cows fathered by proven AI bulls and 48 cows fathered by AI young sires are purchased for a dairy, what is the estimated increase in lifetime profits as compared to heifers purchased with unknown parentage? Assume that the USDA 2001 Animal Improvement Program calculation amounts (see Table 12-2, page 193) are still valid.

REFERENCES

Bewley, J., Palmer, R. W., & Jackson-Smith, D. B. (2001). An overview of Wisconsin dairy farmers who modernized their operations. *Journal of Dairy Science, 84,* 717–729.

Holstein Association. (2001). Cattle merchandising and advertising policy. [Handout]. Brattleboro, VT: Author.

Lawlor, T. (1996, September 19). Study examines value of identified cattle. *Holstein News.*

McCarry, M. R. (1997, January 10). I.D. means identification . . . and spells profit. *Hoard's Dairyman,* 16.

USDA Animal Improvement Programs Laboratory. (2001). Sire evaluation results. Retrieved February 27, 2001, from http://www.aipl.arsusda.gov

Smith, L., & Cassell, B. (1998). The effects of inbreeding on the lifetime performance of dairy cattle. *Journal of Dairy Science, 81,* 2729–2737.

Chapter 13

Heifer Raising Options

KEY TERMS

autopsy
contagious bug
custom heifer raiser
heifer
individual calf hutch
milk pasteurization
mound system
open lot
stall-barn
steam-up
super-hutch

OBJECTIVES

After completing the study of this chapter, you should be able to

- describe appropriate housing facilities for newborn calves, pre- and post-weaning calves, and breeding and pre-calving heifers.
- understand the costs associated with constructing such facilities.
- understand the costs associated with custom heifer raising.
- decide if heifer raising is a financially viable option for an existing dairy, at present and in the future.

Producers have several options for raising **heifers.** During an expansion phase, owning and raising heifers may require resources that would be better spent on the milking herd. Later, as an operation has stabilized and resources become available, owning or raising replacements can become more attractive. The decision to own or not should be based on financial position: can the operation afford to have money invested in heifers? The decision to raise heifers or to contract with a **custom heifer raiser** should be based on the availability of labor, housing, feed, and manure disposal rights needed to properly care for the animals. In addition to these factors, the producer must also consider the biosecurity risks of the different heifer-raising options.

Heifer-Raising Objectives

Every herd needs animals to replace those leaving the dairy. Even under good management conditions, herds that fill with purchased heifers normally experience a 20–30 percent culling rate the first year, sometimes higher. Buying older cows to fill herds can result in high culling rates, especially if the animals are taken from a **stall-barn** environment. The stress of competing in a new environment and the lack of attention by owners can contribute to a high culling rate. Herds with high culling rates need more replacements each year; a shortage of replacement animals has inflated the current cost of replacements. Increased demand, additionally spurred by owners becoming more informed about the cost of raising heifers, has led to the increase in cost of replacements. Knowing what options are available and the costs associated with each can help you determine whether to buy your replacements, have them custom raised, or raise them yourself.

The objective of a good heifer-raising program is to develop healthy, well-grown heifers ready to enter the milking herd at 23–24 months of age. Holstein heifers should be grown to weigh 1,350–1,400 lbs, have a wither height of 53–54 inches, and a body condition score of 3.0–3.5 before calving, and to weigh at least 1,200 lbs after calving. Excessively fat heifers should be avoided since the extra fat can influence udder development and generate calving related problems. The Nutrient Requirements of Dairy Cattle (NRC, 2001) recommends feeding heifers to achieve a growth rate that will support a target weight at first breeding of 52.8 percent of mature weight (55 percent of shrunk body weight, which is 96 percent of full body weight).

Baby calves should initially be housed individually, and then put into small groups to learn to socialize and compete. Calves should be separated from their dams soon after birth and fed colostrum to effect antibody transfer from the dams. Baby calves should be housed away from mature animals to minimize exposure to pathogens. Avoid exposure of calves to adult animal manure. Calf housing should prevent physical contact between animals, to prevent disease transmission and sucking. Housing calves at least six feet apart can minimize airborne transmission of disease. Later, heifers should be grouped by age and size, to facilitate feeding to meet their nutritional needs and to avoid domination by larger animals. Heifer housing should include provisions for individual housing of calves, small socialization training groups, two open-heifer groups, a breeding group, and a bred-heifer group. Heifers should be moved to a pre-calving (**steam-up**) group four to six weeks before calving.

The following lists provide some of the elements of a good heifer-raising program, which must be considered no matter how many replacements will be raised.

For all animals, provide

- a clean, dry, comfortable resting surface.
- a labor-efficient facility designed to support easy cleaning, bedding, and feeding.
- a well-ventilated, draft-free facility designed to prevent dead spots.
- adequate space per animal and sufficient access to feed and water.
- biosecurity measures to minimize risk of exposure to diseases.
- a vaccination program designed to protect animals at current and future locations.
- adequate fly control procedures.
- facility designs that support worker comfort.
- buildings and feeding areas oriented to make the best use of local weather conditions.
- easily maintained animal-identification and record-keeping system.
- adequate shade and shelter, if animals are housed outside.

For heifers, provide

- an easily accessible treatment area designed to minimize animal stress and to support the safety of workers and animals.
- a labor-efficient method for monitoring animal weight and wither height.
- a well-ventilated, draft-free facility designed to prevent dead spots, and outside areas that allow animals to avoid muddy conditions.
- access and time to learn to use self-locking manger stalls or freestalls if animals will be exposed to these after calving.
- an internal and external parasite-control program.

For calves in the maternity area, implement procedures to

- feed colostrum within 30 minutes of birth.
- identify calves, dip navels, and so on, soon after birth.
- separate calves from their dams within one hour of birth.
- place calves in a warming box if born during cold weather.
- protect calves from exposure to adult animals or their manure.

For calves in pre-weaning, provide facilities and procedures that

- housing near a well-traveled area, to allow frequent observation.
- an area to prepare and store feed and supplies.
- the ability to clean and sanitize calf-feeding equipment.
- access to starter and water, starting the day after birth.
- individual pens that can be disinfected easily between calves.
- feed other than hay for the first two months.
- **milk pasteurization,** if waste milk will be fed.
- a calf health program to control coccidian and other health concerns.

For calves in post-weaning, provide facilities and procedures that

- allow calves to build immunity and learn to socialize and compete.
- house five to eight animals per group.
- allow calves of similar size to be housed together.
- allow no more than two months' age difference for animals housed together.

For breeding-age heifers, provide

- an easy method to restrain and artificially inseminate animals.
- a procedure for insertion of rumen magnet.

For pre-calving heifers, provide

- housing similar to milking-cow housing.
- rations formulated to support transition to the expected post-calving ration.
- hoof trimming, and foot baths if needed.

Heifer-Raising Costs

The cost to raise a heifer varies considerably from farm to farm, and region to region. Total cost must include the costs of feed, labor, housing, death losses, breeding, health, hauling, plus any other relevant investment costs. The cost also varies by the size and age of animals. Hoffman (1996) calculated the average total cost to raise heifers from birth to calving ($1,130) and the cost for each 100-lb increment of weight. He documented the average cost per pound and the cost per day for each of these increments. Table 13-1 shows the average feed, labor, and total cost for each increment. These values are very helpful for producers who are thinking about contracting with custom heifer raisers or need to estimate their own cost to raise animals to a specific age or weight.

TABLE 13-1 Cost per day and cost per pound gain, by animal size, for raising heifers.

Start Wt.	End Wt.	Feed $/hd/day	Labor $/hd/day	Total $/hd/day	Feed per lb gain	Labor per lb gain	Cost per lb gain
100	200	1.15	.72	2.54	.69	.43	1.52
200	300	.41	.11	1.06	.25	.07	.64
300	400	.46	.11	1.13	.28	.07	.68
400	500	.49	.11	1.16	.29	.07	.70
500	600	.50	.11	1.22	.30	.07	.73
600	700	.56	.11	1.28	.34	.07	.77
700	800	.63	.12	1.37	.38	.07	.82
800	900	.71	.15	1.76	.43	.09	1.05
900	1,000	.78	.13	1.79	.47	.08	1.07
1,000	1,100	.86	.11	1.72	.52	.07	1.03
1,100	1,200	.97	.11	1.77	.58	.07	1.06
1,200	1,300	1.11	.16	2.06	.67	.10	1.24
Average		.72	.17	1.57	.43	.10	.94

Custom Heifer Raisers

Many producers, in expanding their herds, contract with custom heifer raisers to raise some or all of their heifers. This is can be financially beneficial when heifer housing is inadequate and existing financial resources are needed for herd expansion. Having heifers custom raised may allow additional cows to be milked using available on-farm labor, housing, feed, and manure disposal rights resources, but producers relinquish some of the control of the animals. Producers who choose this option also lose the opportunity to feed their poorer-quality feed to the heifers. Producers often question whether custom heifer raisers can raise animals as well as they themselves can. Table 13-2 contains research results showing that custom heifer raisers can produce heifers at lower costs than on-farm raising (Gabler, 1999). This makes sense, since producers who specialize often achieve economies of scale not realizable with smaller operations. They can afford to implement procedures and monitoring systems to control costs and ensure performance.

Producers considering having their heifers custom raised should research existing growers and determine if they can produce the quality of replacement needed. Ensuring heifers are returned at the proper size, body condition, and calving age is very important. Many growers are members of the Professional Dairy Heifer Growers Association, who may be able to

TABLE 13-2 On-farm versus custom cost to raise heifers.

	On-farm	Custom
Age at calving (months)	25	23
Total cost/hd	$1,128	$964
Daily cost/hd	$1.55	$1.46

supply the names of local growers. Ask other producers about their experiences with local suppliers. Interview growers, inspect their facilities, and ask for their rates. Always make a written contract that clearly states the performance standards and specifies who will pay the costs of hauling, insurance, breeding, dehorning, hoof trimming, vaccination, veterinary services, **autopsy,** and animal identification. Liability for death losses should be clearly spelled out.

Heifer Facilities

Ideal heifer-raising facilities would allow individual housing of baby calves, small-group housing for them to learn to socialize and compete with minimal stress, and then larger-group housing, which is more labor efficient and allows the animals to learn to compete when they enter the milking herd. Many different designs have been tried, with varying success; facility management is a significant factor as well. Old buildings can be successfully converted and work well if ventilation and sanitation are adequate. New facilities often are more labor efficient but may have higher initial costs.

Calf Housing

Research has shown that calves raised in cold housing grow faster and have fewer health problems than those raised in warm housing. Outdoor **individual calf hutches** perform well but can be difficult to manage during bad weather (Figure 13-1). Worker discomfort, manure handling difficulties, snow buildup, freezing of water, and need to provide summer shade are common complaints. Moving hutches under the cover of a building increases cost but can alleviate some of these concerns. Hutches should be placed at least six feet apart to prevent the spread of airborne disease and should face south to protect calves from the north wind. Calves should be tied or given a small exercise yard outside the hutch. To better utilize space and cut cost of construction, barns with individual pens have been used successfully. This option increases initial cost but solves some of the problems associated with outdoor hutches. Since animal density (and thus proximity) is increased, care must be taken to properly sanitize stalls between calves,

FIGURE 13-1 Individual hutches can be used very effectively to raise healthy calves.

FIGURE 13-2 Curtain-sided calf barns, with individual calf pens, can be used successfully if ventilation and sanitation is properly controlled.

and to monitor animal health in case a **contagious bug** gets introduced into the facility. Figures 13-2 through 13-4 show such a complex. The curtain sidewalls allow the barn to be opened in the summer (to provide shade and proper airflow), and closed as weather conditions change (to prevent drafts). The individual pen dividers extend past the front of the pen and prevent calf contact. These individual dividers can be removed to create small group pens that allow several calves to socialize and learn to compete without the stress of being moved. Later, more dividers can be removed to make even larger pens, or the animals can be moved to group pens.

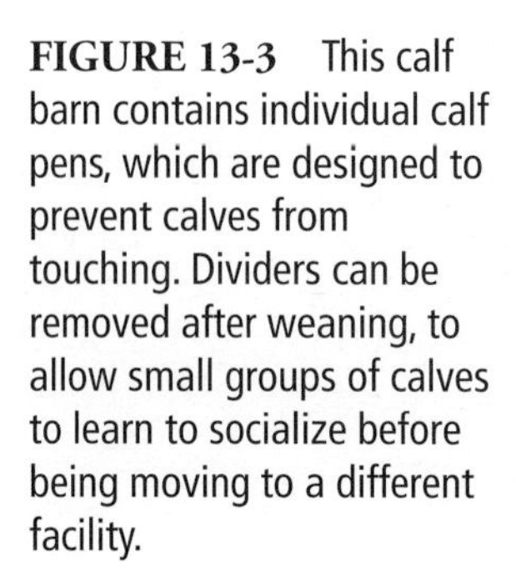

FIGURE 13-3 This calf barn contains individual calf pens, which are designed to prevent calves from touching. Dividers can be removed after weaning, to allow small groups of calves to learn to socialize before being moving to a different facility.

FIGURE 13-4 Individual calf pens such as those shown can be easily sanitized.

Post-Weaning Calf Housing

After calves are weaned and are about six to eight weeks old, they should be moved to small pens containing five to eight animals and allowed to learn to socialize and compete. Success has been reported with **super-hutches** (Figure 13-5) and small pens within heifer-rearing barns (Figure 13-6). Ventilation, feeding, and manure management must be considered when making this choice. Super-hutches are normally built on skids that allow them to be moved periodically. Group pens in existing barns can be used if there is proper ventilation (Figure 13-7).

Heifer Housing

Heifers should be housed in groups, based on their age and size. Space requirements change as animals grow, and neck rail heights, curb heights,

FIGURE 13-5 Super-hutches allow small groups of young animals to learn to socialize and compete.

FIGURE 13-6 Small pens can be placed at the end of heifer barns to house weaned calves.

and stall sizes change. Total expected herd size should be considered when a facility is designed, and pens should be sized to support the number of animals of each size expected. Normal recommendations call for younger animals to be housed on bedded packs. After animals reach six to eight months of age, bedded-pack or freestall facilities are recommended. Larger animals can be successfully raised on **open lots** or **mound systems** (Figure 13-8), which

FIGURE 13-7 Conversion of old dairy barns to house calves allows use of existing buildings and manure handling systems.

FIGURE 13-8 Heifer mounds have proven to be an efficient way to house older heifers. Correct mound slopes and proper drainage patterns are important to avoid muddy conditions.

normally are less expensive to build. To ensure proper growth and efficiency, muddy conditions should be avoided. Heifer barns with pens (Figure 13-9) and freestalls (Figure 13-10) sized to match heifer sizes work well. Drive-by feeding systems with TMR minimize the labor of feeding and support proper animal nutrition.

FIGURE 13-9 Modern heifer barns allow several groups of heifers to be fed a TMR efficiently.

FIGURE 13-10 Housing heifers in freestalls minimizes bedding cost, reduces labor requirements, and allows animals to be trained to use freestalls before entering the milking herd.

SUMMARY

Expanding the dairy herd has major implications for the size of the replacement herd needed to support the expanded dairy. If replacements are raised, it will take a minimum of two years before the number of replacements is sufficient to meet the expanded dairy's needs. Heifer housing and associated feed requirements constantly increase over this time, as additional heifers are being raised. Producers may sell calves and buy replacements, hire someone to raise replacements, raise them themselves, or a choose combination of these options. Producers planning to raise their replacements must remember that calves from purchased heifers are often small and difficult to raise or may be from undesirable sires. During the expansion phase, financial resources are often limited, so raising heifers may not be feasible. Later, after the herd has stabilized and resources are available, raising heifers on-farm may allow the producer more control and minimize biosecurity concerns.

CHAPTER REVIEW

1. What factors most contribute to high culling rates for older cows?
2. Explain why is it important for a calf to be socialized in its first months of life.
3. List three characteristics of a good pre-weaning calf facility, and three characteristics of a good post-weaning calf facility.
4. List three provisions that must be made for heifers as they are growing.
5. What is the primary benefit of custom heifer raising? What is the primary drawback?
6. Using Hoffman's figures, how much will the feed cost to raise a heifer from a weight of 400 lbs to a weight of 1,000 lbs? Assume the price of feed per cow per day is $0.62 on average, and that the cow will grow 1.8 lbs per day.

REFERENCES

Dairy freestall housing and equipment handbook (7th ed.). (2000). [Publication MWPS-7]. Midwest Plan Service.

Gabler, M. (1999, July). Developing the next generation. *Midwest Dairy Business,* 20.

Hoffman, Patrick C. (1996). Custom raising heifers: Variation in the cost of raising replacement heifers. [Handout]. Madison: University of Wisconsin, Department of Dairy Science.

National Research Council. *Nutrient requirements of dairy cattle* (7th rev. ed.). (2001). Washington, DC: National Academy.

Chapter 14

Labor Requirements and Scheduling

KEY TERMS

benchmark database
compensation
cow pusher
estrus
FTE (full-time equivalent)
job description
milking herd enterprise
milking shift
organizational structure
owner's draw

OBJECTIVES

After completing the study of this chapter, you should be able to

- list the types of employees required to appropriately staff an existing dairy during its expansion.
- create a plausible work schedule for each phase of the dairy's expansion that accurately predicts its staffing requirements.
- list the compensation required by each type of employee, including nonmonetary forms of compensation.
- develop a labor budget for a dairy expansion.

During the planning process for an expanding dairy you must consider not only facility, equipment, and replacement animals, but also labor needs, employee scheduling, and labor management. The number of employees, the types of employees, and realistic **compensation** costs must be estimated.

Number of Employees

The size of the operation, geographical location, number of different enterprises involved, and amount of work that will be contracted all impact the labor needs of a dairy. **Benchmark databases** from your area can be used to develop a rough estimate of the labor needs of a dairy. These must be used cautiously because they represent an average of dairies that may not be similar to your planned operation. Cows per worker, milk per worker, and labor costs per hundredweight of milk produced can be used to estimate the number of **FTEs (full-time equivalents)** needed to manage the proposed dairy.

Planning for a modern dairy takes a great deal of time, to ensure that each element of the design is correct and compatible with other system elements. Existing dairy producers must understand this and arrange for additional support during the planning and implementation processes. Producers frequently underestimate this, and either the existing operation or the planning process suffers (Figure 14-1). The start-up of a modern dairy often

FIGURE 14-1 Producers must consider the time needed to monitor the construction process, and budget for additional worker and management needs during this time.

FIGURE 14-2 Breaking newly freshened heifers to use a new parlor requires additional labor, which should be considered when planning a major dairy expansion.

requires calving many new animals and training cows to use a new parlor (Figure 14-2). Money should be budgeted for additional management and workers, during this transition period, which often lasts 6 to 12 months. If several new employees are to be added, it is wise to hire them before the start-up so they can be familiar with both facility and procedures. (Training employees on the computer systems to be used would be a huge advantage.) Remember that not all employees will remain with the operation, and it is easier to terminate someone after completing the start-up process than to hire people during this phase or to go through the process with insufficient labor (this is true of both workers and management). Do not overestimate the abilities of yourself and your team, or you may create a start-up nightmare (Figure 14-3).

The number of hours each worker type will work per week is influenced by job type, shift length, herd size, and type of employee hired. Results of an April 2000 survey of Midwestern and Western dairy producers showed that among milkers and **cow pushers** women averaged 36.8 hours per week and men averaged 45.7 hours per week, whereas herd managers averaged 10.5 hours per day, feeders averaged 8.8 hours per day, and other full-time employees averaged 8.2 hours per day (Bower-Spence, 2000). The survey showed the average wage of dairy herd managers to be $10.30 per hour, and cow feeders and all-around employees $9.20 per hour.

The results of the 1999 Wisconsin Dairy Modernization Survey (Palmer & Bewley, 2000) show the number of employees in 1999, by

FIGURE 14-3 Managers must allocate and budget for adequate time to perform their management functions.

TABLE 14-1 Labor usage by size of herd (Palmer & Bewley, 2000).

Labor-Related Factors	60–105	106–145	146–220	221–360	>360
Number of family members	2.93	3.10	3.20	3.28	3.52
Number of full-time employees	1.29	1.38	2.03	2.62	6.89
Number of part-time employees	1.48	1.84	2.91	3.18	4.85
Total hours per person per week	52	48	46	46	48
Yearly hours per cow	111[a]	84[b]	72[c]	60[d]	56[d]
Cows per full-time equivalent	27[c]	34[b]	40[b]	49[a]	51[a]
Acres per cow	3.3[a]	3.3[a]	2.6[b]	2.6[b]	2.3[b]

[a–d]Within rows, superscripts represent significant differences.

employee type, for producers who expanded their dairies between 1994 and 1999 (Table 14-1). Larger herds added more full-time and part-time labor, but there was no significant difference between herds in hours worked per person per week (46–52). Farms with larger herds appear to be achieving greater labor efficiency, since yearly hours per cow decreases from 111 for smaller herds to 56 for larger herds, and cows per full-time equivalent increased from 27 to 51 cows. Herds with 221–360 cows, and more than 360 cows, had significantly higher numbers of cows per FTE and lower yearly hours per cow than other size groups. (FTEs were calculated for this survey using 50 hours per week. If you are using FTE or benchmark numbers to calculate the number of employees needed, remember to convert FTEs using the expected average shift length for your dairy—for example, 12 employees working 46 hours per week = 10 FTE). Part of the difference in the overall labor efficiency of farms may be due to the amount of cropping done by each group. Table 14-1 shows that the average number of acres per cow is significantly smaller for larger herds.

Using these values, a Midwestern dairy should probably plan for 60–75 cows per FTE for a **milking herd enterprise** only, 50–60 cows per FTE for a milking herd plus a heifer or crop enterprise, or 40–50 cows per FTE if milking herd, crop, and heifer enterprises are included. Large herds in the West often achieve 100 cows per FTE or greater, but they purchase their forage, have their heifers custom raised, have cows in **estrus** identified and bred by a custom breeder, and have their financial records maintained by private contractors. Dependence on outside services decreases the number of on-farm workers needed, but the cost of these services must be considered elsewhere in your financial analysis.

Compensation

Employee compensation can come in many different forms. Cash wages, bonuses, and incentive payments must be considered, along with compulsory costs such as social security, workman's compensation, and unemployment insurance. Other benefits (such as use of a house, food, time off, and other insurances) may also be provided.

To calculate the expected labor cost of a larger dairy, you must first determine its **organizational structure** and staffing requirements. Will the herd be milked two or three times per day? How many people will be needed to run the milking parlor based on your milking procedure? Will crops or heifers be raised? An organizational chart should be developed that lists the roles and reporting order of each team member. How many employees total, how many of each type, and fair compensation for each (based on local conditions and competition for labor) must then be determined.

Table 14-2 shows the average wages reported in 1999 by survey respondents for different employee classifications. Full-time herd managers were the highest-paid employees, and established employees were paid more

TABLE 14-2 Average wages of dairy employees (Palmer & Bewley, 2000).

	New Employees				Established Employees			
	$/hr	No.	$/mo	No.	$/hr	No.	$/mo	No.
Managers (full time)	8.48	22	2,275	34	10.58	21	2,307	54
Nonmilkers (full time)	7.18	76	1,762	21	8.53	77	2,019	27
Milkers (full time)	7.32	131	1,596	16	8.87	132	1,779	29
Milkers (part time)	6.80	162	750	4	7.96	147	984	8
Other (part time)	6.37	113	963	4	7.59	104	950	4

than new employees. Managers tended to be paid monthly salaries rather than hourly wages like other job classifications.

Using your own experience, information from other local producers, and survey information such as this, wage rates can be established. To calculate the total labor cost, nonwage costs must be added: Social Security is normally 15.3 percent of wages, workman's compensation and unemployment insurance is based on the number of employees expected, and the types of other benefits and their costs must be determined. Table 14-3 shows the types and frequencies of benefits provided to employees of surveyed herds. The most common benefits are paid vacations and health insurance.

After you have determined the number and type of employees and your expected total cost of labor, check to ensure that these values are reasonable. Table 14-4 shows actual total labor cost per cow and per cwt of milk in 1999, without **owner's draw,** as reported by dairy farms in

TABLE 14-3 Employee benefits (Palmer & Bewley, 2000).

Benefits Provided to Full-Time Employees	Number of Herds
Paid vacation time	144
Health insurance	143
Milk or meat	107
Housing	89
Other noncash benefits	38
Profit-sharing	24
Allow employee to own animals in herd	20
Retirement plan	19
Share of calves born	7

TABLE 14-4 Total labor costs (Palmer & Bewley, 2000).

	Herd Size			
	76–100	101–150	151–250	>250
Average number of cows/herd	87	122	194	443
Milk shipped/cow	20,039	20,479	19,979	21,693
Total labor cost w/o owner draw	$29,648	$44,080	$73,037	$223,502
Total labor cost/cow/yr w/o owner draw	$341	$361	$376	$505
Total cost/cwt shipped w/o owner draw	$1.70	$1.76	$1.88	$2.33
Benefits as percentage of wages	35%	27%	23%	21%

Wisconsin. Remember that as dairies get larger, their organizational structure often changes, and an owner's draw is often treated as a labor wage and included in total labor cost. Large, well-run dairies can expect an average labor cost per hundredweight of $1.75–2.25. A goal of a million pounds of milk per worker is often mentioned, but this amount is greatly exceeded by operations when only the milking herd enterprise is considered.

Scheduling Labor

Having a feeling for how labor will be scheduled can be valuable during the planning process to estimate the number of employees needed, but the estimate may change as circumstances change. Many dairies have experienced scheduling changes after modernization, as they changed employee types. One dairy moved the **milking shifts** to 9:00 A.M. to 5:00 P.M. and 9:00 P.M. to 5:00 A.M. to accommodate working mothers who needed to conform to the schedules of day care centers. Some immigrant groups like to work long hours (often 12-hour shifts) and long work weeks. Since agriculture is exempt from overtime pay requirements, employees putting in long hours are not a concern as long as their safety or productivity is not affected.

A common practice of large dairies has been to establish eight to nine hour shifts. Workers often overlap shifts by a few minutes to exchange information. If mainly full-time employees are to be used, a work schedule different from seven days per cycle allows workers to alternate days off so everyone gets an equal number of weekend days off.

Herds milked 3X often hire four milkers and have a schedule of six days on and two days off (six-on, two-off). This allows 24 milkings in the eight-day rotation to be evenly distributed between four people, giving each person six milkings per rotation. Herds milked 2X often hire three milkers and select the four-on, two-off rotation, giving each milker four milkings in the six-day rotation. If your dairy requires two milkers and one cow pusher per shift, 12 full-time employees will be needed. When determining the number of employees needed to perform the milking process, remember that the expected milking procedure will impact the shift length and/or the number of employees needed (Figure 14-4). If a full-prep milking procedure is to be used, plan for one milker per 8–10 stalls on a side of a parallel or herringbone parlor.

Here is an example of a 3X milking rotation with four milkers (milkers *1*, *2*, *3*, and relief milker *R*), in a six-on, two-off schedule.

Day								
Milking	1	2	3	4	5	6	7	8
A.M.	1	1	1	1	1	1	R	R
P.M.	2	2	2	2	R	R	2	2
3X	3	3	R	R	3	3	3	3

FIGURE 14-4 Parlor type and milking procedures impact the labor requirements of a dairy.

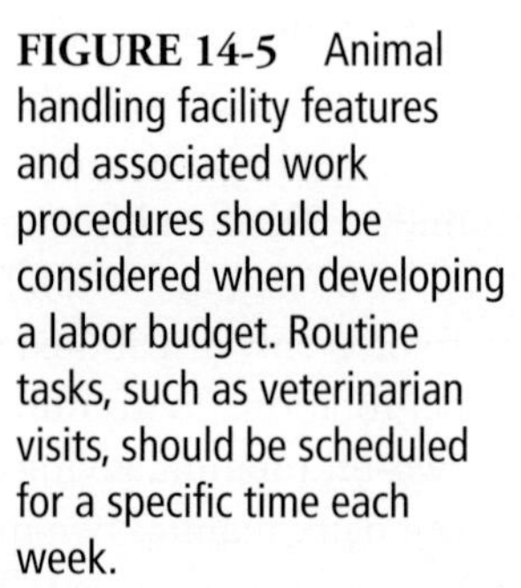

FIGURE 14-5 Animal handling facility features and associated work procedures should be considered when developing a labor budget. Routine tasks, such as veterinarian visits, should be scheduled for a specific time each week.

If this type of schedule is used, the weekly equivalent hours can be calculated as follows:

$$\text{weekly equivalent (hours)} = \frac{\text{days worked} * \text{hours/day}}{\text{days in rotation}} * 7$$

For example, for a worker on an eight-day rotation (six-on, two-off), working nine hours per day: $(6 * 9)/8 * 7 = 47.25$ hours.

To minimize the number of employees needed on weekends and holidays, and to facilitate coordination with off-farm support people, most large dairies have developed a schedule of routine activities (Figure 14-5). This standardization of routines ensures that all tasks are accomplished and enables workers and support people to plan their schedules. An example of such a schedule follows.

Monday	Breed heat-synchronized cows
Tuesday	Veterinary health and reproductive examinations
Wednesday	Dry-off cows and move animals
Thursday	Bed stalls and clean bedding packs
Friday	Prepare for weekend and next week

Types of Employees

The work performed on all dairies is very similar, but a larger dairy can allow an employee to do fewer tasks than can a smaller dairy. This employee specialization decreases the time needed to train new employees and achieves a higher competence level with the assigned activity. Organizing compatible functions into specific jobs, locating them in your organizational chart, and writing meaningful **job descriptions** should enable you to determine the types of workers needed. Jobs seen on large dairies include dairy manager, herdsperson, assistant herdsperson, head milker, milker, cow pusher, fresh-cow manager, hospital manager, breeder, crop manager, field worker, replacement manager, calf feeder, and so on. Larger herds should consider hiring a receptionist to support management (Figure 14-6). After each job has been defined, you should specify the knowledge, skills, education, and experience needed. In addition to these formal requirements, you may search for employees who have an attitude that complements your management style and who will work well with the remainder of your workforce.

FIGURE 14-6 Modern dairies require extensive records, and the cost of supplies and office-support personnel should be considered in labor and financial budgets.

SUMMARY

When planning a new or expanded dairy, you must accurately determine your labor requirements, compensation costs, and labor management needs if the dairy is to be successful and meet financial projections. Remember that the planning and the start-up stages require extra worker and management efforts that must be considered and budgeted for.

CHAPTER REVIEW

1. Explain the advantage of hiring employees before expansion.
2. What four factors determine the number of hours each employee type will work?
3. What is a full-time equivalent (FTE), and what is it typically equal to?
4. List five monetary forms of compensation often provided to employees; then list five nonmonetary forms.
5. What aspects of modernization typically require additional labor? For each aspect, develop a list of employee types that must be hired.
6. Create a plausible operational schedule for a dairy that takes into account milking shifts, tasks to be completed, the number of employees on the job, and the lengths of workers' shifts.

REFERENCES

Bewley, J., Palmer, R. W., & Jackson-Smith, D. B. (2001a). Modeling milk production and labor efficiency in modernized Wisconsin dairy herds. *Journal of Dairy Science, 84,* 705–716.

Bewley, J., Palmer, R. W., & Jackson-Smith, D. B. (2001b). An overview of Wisconsin dairy farmers who modernized their operations. *Journal of Dairy Science, 84,* 717–729.

Bower-Spence, K. (2000, November/December). Wage survey results. *Dairy Today,* 33.

Frank, G., & Vanderlin, J. (2000). *Milk production costs in 1999 on selected Wisconsin dairy farms.* Madison: University of Wisconsin, Center for Dairy Profitability.

Palmer, R. W., & Bewley, J. (2000). *The 1999 Wisconsin dairy modernization project—Final results report.* Madison: University of Wisconsin.

Chapter 15

Labor Management

KEY TERMS

affiliative
authoritative
coaching
coercive
delegating
democratic
employee handbook
employment-at-will
job requirement
job summary
mission statement
open-ended question
operating procedure
pacesetting
recruitment

Objectives

After completing the study of this chapter, you should be able to

- identify common leadership styles and the stages of employee growth.
- explain the function and components of an organizational chart.
- understand the legal significance of a company handbook, employee job descriptions, appropriate hiring techniques, and employee evaluations.
- understand the benefits of marketing to prospective employees.
- develop a marketing, hiring, training, and employee evaluation plan for an existing dairy.

The most critical aspect of expansion for many traditional dairy managers is the change in role from worker to manager. Job descriptions, **employee handbooks,** operating procedures, and organizational charts for the dairy operation should help. Finding, hiring, training, and motivating employees to satisfy the increasing labor needs of the dairy is a challenge. Managing employees in a way that causes them to follow proper milking procedures, maintain a sound animal health program, and emphasize milk quality can contribute substantially to the bottom line. Producers planning a major expansion or planning a new large dairy must develop personnel management tools *before* employees are hired and cattle arrive.

Leadership Styles

As the manager of a modern dairy, you will need to guide the activities of people working for you in the pursuit of a common set of goals. People have default leadership styles that are natural to them. This natural leadership style is used without thinking; though it can be adjusted with training, it normally reappears under stress. Successful leaders must adapt their leadership styles to employees' individual stages of development. Tests have been developed to identify natural leadership style; if you plan to manage people, it would be advisable to take one. Using the results of such a test, you can analyze your leadership style and start to understand what you must do to manage different types of employees. One such test, the McClelland Leadership Style Questionnaire, defines six leadership styles: **coercive, authoritative, affiliative, democratic, pacesetting,** and **coaching** (AgVentures, 1998).

Coercive—expect immediate compliance; "do it or else"

Authoritative—manage without doubt of who the boss is; "firm, but fair"

Affiliative—people come first and tasks second; "good buddy"

Democratic—have a participative style; "let's vote"

Pacesetting—perform technical activities as well as manage; "follow me"

Coaching—develop subordinates; "developer, delegator"

No single leadership style is best, because each has its advantages and disadvantages. It is important to understand your natural style and learn to modify your behavior depending on the circumstances. Managers who do not understand their management style and do not modify their behavior based on individual employee differences, often have excessive employee turnover (unless or until a team of individuals who happen to like this management style is found).

Employee Growth

Situational leadership is defined as adapting one's leadership style to the needs of the employee, according to the employee's individual stage of development (AgVentures, 1998). In this schema, four steps lead an employee to a position of authority. This concept is important because keeping good employees often is associated with allowing them to grow within the organization. If any of these steps are skipped or passed through too quickly, problems will arise.

Step 1. Directing—there is one-way communication between you, the teacher, and the employee, who is new and does not know what to do.

Step 2. Coaching—as the employee grows in competence and confidence, you begin to involve him or her in decisions.

Step 3. Supporting—the employee shoulders most of the responsibility, and you are available for support if needed.

Step 4. **Delegating**—the employee has the knowledge and confidence to take on the responsibility completely, and the manager's role is to evaluate his or her performance.

Using this model of leadership, the manager must understand the employee's status and personality, and both motivate the employee and help him or her grow within the organization.

Organizing the Workforce

All dairy farms with multiple workers have some type of reporting structure, formal (written) or informal. A formal organizational chart allows the farm manager to define each team member's role and how the tasks fit together to achieve the organization's goals. Roles are grouped according to sections of the business and show reporting paths for each employee type. A job description should be generated for each position in the organizational chart. All the tasks that must be done on the farm should be included in the job descriptions.

An organizational chart shows the design of the organization's structure (Figure 15-1). It should provide a map that lets each team member know which supervisor to contact in a decision or problem situation. Authority and responsibility should be delegated to the lowest level possible, to free managers from making basic decisions and allow them to accomplish their work. A manager can normally supervise five to six employees, but this number can vary depending on the complexity and variety of the jobs involved. Organizations should be structured so that each person has only one supervisor. Otherwise, conflicting instructions and assignments may confuse the employee.

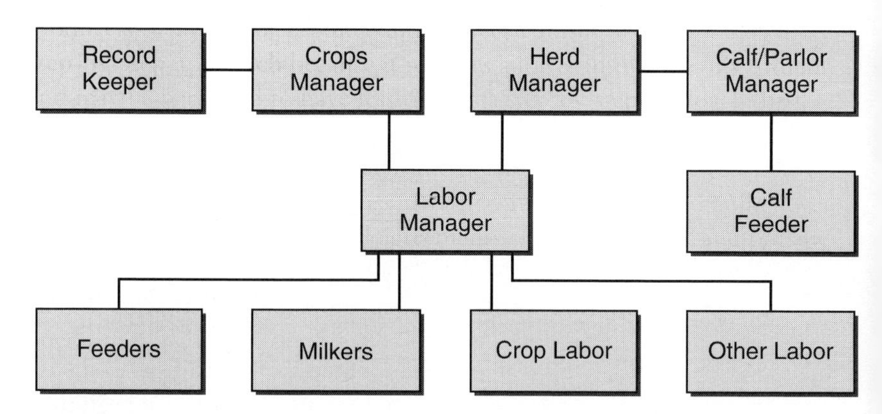

FIGURE 15-1 This farm family hired a labor manager to direct the activities of all hired employees, while the mother did records, the father was in charge of the cropping operation, the son in charge of the dairy, and the daughter-in-law in charge of calf management.

Job Descriptions

There should be a job description associated with every role listed on the organizational chart. The chart and job descriptions should be created early in the planning process and used for planning of labor requirements and hiring of new employees. Job descriptions should include a job title that both explains the nature of the job and has a positive connotation, along with the following elements: a **job summary,** which briefly states major responsibilities; **job requirements,** which state the knowledge, skills, and abilities required (make sure these requirements are essential to perform the tasks and do not violate any civil rights); scope of tasks and duties, which lists major activities performed; reporting relationships, which includes both whom the employee reports to and who he or she supervises; and performance standards and expectations, which describes what is expected of the employee.

Operating procedures should also be referenced in the job description. They should be very detailed step-by-step instructions that explain how a task is to be performed. In the planning phase of a dairy, before it is actually operating, these documents may be general, but once the operation is functioning and procedures refined, they should contain detailed instructions that are updated as procedures change. They should be included in the employee handbook, with the job descriptions that refer to them, and often will be posted at the site of the activity. (For example, a milker's job description may state that one of his or her duties is to milk cows following the dairy's established milking protocol, which is defined in the dairy's "milking procedure.")

Once job descriptions are created, they can serve many different functions, as in the following examples.

Recruitment
  Prepare help-wanted adds
Selection
  Screen applicants
  Select the best applicant
Training
  Orient new employees
  Train new employees
  Develop mutual understanding between employee and employer
Performance appraisals
  Develop performance standards
  Employee evaluation
  Job recognition
  Corrective action
Compensation management
  Wage increases
  Basis for bonus or incentive payments
Corrective action
  Dismissal
  Defense, if a dismissal is challenged

Keep job descriptions brief by leaving out unnecessary details. Stress *responsibilities* for tasks, rather than the actions themselves. Divide farm tasks across job descriptions to offer some variety. Remember to update job descriptions as roles of employees change.

Employee Handbook Development

An employee handbook is a written document that tells employees how you run your business, what you expect of them, and what they can expect of you. It is an educational document that informs each employee about your business's policies and procedures. It should contain a copy of the farm **mission statement** (Figure 15-2 for an example), an organizational chart, a map of the facilities, company policies, employee job descriptions, and operational procedures. If correctly written, it can reduce communication problems and conflicts and save time in training employees. Such a document also tends to present a good image for your business and attracts high-quality employees. It is an excellent way to consistently share the information needed by every employee. It demonstrates that you have thought through your policies and allows you to ensure that everyone has been informed. Each new employee should receive a copy of an employee handbook as well as a copy of procedures specific to his or her job.

Mission Statement

- A mission statement should clearly define your organization's goals and objectives.
- Write it, post it, and live it.
- Example:

Best-Dairy-Ever exists to provide
- a good standard of living to its owners and employees;
- a pleasant and safe environment;
- sufficient free time for family, social, religious and other off-farm interests;
- a reasonable return on its owners investment; and
- sufficient profits to improve the business at a rate that will keep it competitive in the dairy industry.

FIGURE 15-2 Every modern dairy should have a mission statement that clearly states the goals of the organization.

The handbook should include information on bonus and incentive programs, trial or probation periods, rules of conduct, disciplinary procedures, time-off policies, and reasons and procedures for termination. Rules relating to harassment, alcohol or drug abuse, smoking, dress code, pay procedures, and insurance or other benefit options should also be stated. Any differences in rules for full and part-time employees should be made clear. Language used in the manual should be clear and straightforward. It should be organized using bulleted points or short sentences rather than long paragraphs. It should emphasize positive "do" rather than negative "do not" statements.

When writing this document, be careful to consider legal implications, since it may be considered an implicit contract, meaning that you could be sued if you do not follow stated procedures. Always use the phrase **"employment-at-will,"** which means that either the employer or employee can terminate employment at any time; failure to include it could result in legal troubles later. Try to use the word "may" instead of "will" to prevent legal problems; for example, "may result in dismissal" instead of "will result in dismissal." Be careful not to include anything that could be considered discriminatory on the basis of race, color, sex, sexual orientation, ancestry, citizenship, age, family status, handicap, marital status, ethnic origin, or place of origin.

Hiring the Right Employees

Recruitment of the right people leads to an efficient and profitable operation, through increased productivity and decreased stress. Hiring the wrong employees can increase stress and decrease efficiency and profit because of extra management efforts and labor turnover. Prospective employees should be told what is expected of them with respect to work ethics, attitude, and

FIGURE 15-3 To attract superior workers, create a positive image of your operation and a safe, attractive work environment.

job priorities. Before hiring an employee, try to determine his or her skills, personality, experience, and attitude. To prepare to attract and hire superior workers, you should create a positive image of your operation using publicity, such as special events ("Breakfast on the Farm," etc.), brochures, or farm tours. Cultivate a professional atmosphere around your operation to give prospective employees the feeling that it is a place where they will enjoy working and can achieve personal growth (Figure 15-3). Employers who treat their employees right often have a waiting list of people wanting to work for them, whereas other producers have high turnover rates and difficulty attracting qualified employees.

The availability of employees varies by geographic area, depending on the competition for labor. A dairy can be staffed by any combination of full- and part-time, male and female, and local and foreign workers. In recent years, a large number of foreign workers, many of them Spanish-speaking, have been welcome additions to the dairy workforce. When workers speak a different language from their manager, every effort must be made to communicate with them. Learning their language, helping them learn your language, translating written materials, and hiring bilingual staff all help. Knowing the cultural differences between members of your workforce will help you prepare for any management problems that may develop.

To continue to attract well-qualified candidates and to avoid legal problems, you must have a well-developed job-application and -selection process. Keep current on recent interpretations of civil rights law, and state and federal employment legislation. Be sure that older applicants, women, minorities, and the disabled are treated in the same way that you treat other people. Follow standard hiring procedures; deviations from them may result in legal and labor disputes. Require everyone to complete a job application form on-site. Do not allow forms to be taken home and dropped off.

Try to advertise in enough locations to get a pool of applicants to choose from. Use the job application and your job descriptions to screen applicants. Select two or three that seem to be the best qualified for the job, and verify their references before inviting them to a formal interview. When reviewing a person's background, you must determine why he or she has changed jobs or has gaps in employment history. The job application form should not pose questions that could be considered discriminatory, and it should have a place where the applicant signs and dates the document.

Interviewing

The interview process gives you the opportunity to evaluate prospective employees, and gives them a positive first impression of your operation. Ask all applicants to complete a job application, develop a list of **open-ended questions** that encourage the applicant to talk, and try to determine if the person is qualified, motivated, and will fit in with your existing team. Select questions that avoid potential legal liability. Do not ask questions that reveal the fact that the applicant is a member of a particular minority, gender, or class. Only ask questions related to the specific job for which the interview is being conducted. Try to determine the stability and dependability of the applicant by reviewing and asking questions about previous work experiences. Since the applicant must work as part of a team, ask questions to determine the applicant's ability to get along with others: Does the candidate speak well of former employers? Does the candidate listen and respond well to questions? Are there signs of defensiveness, aggressiveness, or timidity? After the interview, evaluate the applicant on previously established criteria; take notes of special strengths, weaknesses, and experiences. Do not oversell the job, but give a positive and valid impression of what the role and responsibilities of the job will be.

After you have interviewed the candidates who passed your initial screening and selected the best one, it is time to offer the job. Review with the candidate the job description, your expectations, and the conditions of employment. Encourage questions and clarify job responsibilities. Confirm that the applicant knows about any special job requirements. The offer, for example, may be dependent upon passing a medical exam or not smoking, drinking, or taking illegal drugs on the job. Allow time for the applicant to consider the offer and schedule any medical exams, and for you to prepare hiring documents. Rejected applicants should be notified of your decision, and if you must reject applicants who were seriously considered, you may want to tell them that you will keep their applications on file in case other opportunities materialize. When a candidate accepts a job offer, an employee agreement that clearly states the job responsibilities and terms of employment should be completed, signed, and dated by both you and the employee.

Training

Training of employees includes both the initial orientation of a new employee and the ongoing training that develops the competence and confidence that leads to employee commitment to the organization and the job. During initial training, you must develop communications channels that lead to friendly relations and understanding of the farm's business goals, as well as job details and work rules. Develop a training form that lists all of the things a new employee needs to know and understand. Review the job assignment, the contents of the employee handbook, and job and procedure descriptions. Tour the facility and introduce the new employee to other employees and others with whom he or she will be interacting. Demonstrate how things should be done. Document all training; have the employee sign the completed training document, and keep this document in the employee's personnel folder.

Performance Evaluations

Employees want and need to know how their performances measure up. Formal evaluation is a tool that the producer can use to improve communication and to help develop employee potential. Plan to have informal conferences several times per year, and formal evaluations at least once per year. New employees should be evaluated before the end of their probation period. Evaluations should be based on previously established performance standards; you should compliment the employee on jobs done correctly, and offer suggestions on improving performance when necessary. These meetings can benchmark an employee's training program, serve as a reference point for corrective action, and provide a defense in court if a termination is challenged. Results of the formal evaluation should be in writing and kept in the employee's personnel file.

Conflict is a natural part of life, and usually arises when communications break down or disagreements occur. The key to successful management of a dairy is to identify conflicts early and take corrective action before they escalate or spread to other employees. To successfully manage conflict, the employer must describe the problem in a nonjudgmental way, involve everyone concerned, find a mutually acceptable solution, and create a sense of shared responsibility for the solution. Questions should elicit information, not belittle employees. Parties involved should avoid getting angry or defensive or feeling threatened. All conversations relating to inappropriate behavior should be documented, dated, and kept in the employee's file. A formal plan for disciplinary action should be part of the employee handbook. Repeated occurrences may lead from informal discussions to verbal warnings, to written warnings, to suspension, and finally to dismissal. Remember that the reason for dismissal must be work related, the employee must have been informed of the standards of performance and behavior expected, and he or she must have knowledge of the policies and consequences. The burden of proof that an employee was handled properly rests with the employer, so be careful when administering discipline and document dates, times, and circumstances involved.

SUMMARY

To manage a modern dairy successfully, you must surround yourself with high-quality, motivated employees. As the manager, you must select the right people and establish an environment that allows employees to achieve their personal goals and the goals of the operation. If a major change in herd size or number of employees is being considered, a plan should be developed that identifies the number and type of employees needed as well as the personnel management procedures for hiring and managing them. This plan should be in place before the first additional person is hired. If working with hired labor is new to you, consider taking courses to learn more about your leadership style and about people's different personality types. Labor management sessions and courses are often included in educational programs sponsored by university cooperative extensions, feed companies, and professional organizations. The AgVentures program in Wisconsin (conducted periodically) and the Western Dairy Management Conference in Nevada (held every other year) are good examples of the sort of programs sponsored by cooperative extensions.

CHAPTER REVIEW

1. List the primary characteristics of each of the following leadership types: coercive, authoritative, affiliative, democratic, pacesetting, coaching.
2. How must a leader change in order to accommodate employee growth?
3. List four components of an organizational chart.
4. List six types of information contained in a job description.
5. List ten pieces of information contained in an employee handbook.
6. What type of question is most productive in an interview?
7. Explain the relationship between employee training and employee evaluation.
8. How can language barriers be minimized in training and daily operations?

REFERENCES

AgVentures. (1998). *Human resource management handbook.* Madison: University of Wisconsin, UW Extension and Center for Dairy Profitability.

Ontario Agricultural Human Resource Committee. (1996). *Writing and using your employee handbook for agriculture and horticulture, an employers' guide and work book for farm owners, managers and supervisors.* Calcdonia, Ontario: Author.

Chapter 16

Record-Keeping Systems

KEY TERMS

chart-of-accounts
dairy herd management (DHM) system
dairy record processing center (DRPC)
electronic dairy-data collection (EDDC) system
milk conductivity
milk flow rate
milk-flow meter
peak milk
unit attachment time
unit on-time
vacuum stability

OBJECTIVES

After completing the study of this chapter, you should be able to

- list the types of data that record-keeping systems collect, and explain their use in dairy management.
- understand the benefits that electronic record-keeping systems offer for individual operations and the dairy industry as a whole.
- understand the limitations and potential pitfalls of electronic record-keeping systems.
- identify additional facility expenses associated with electronic production-monitoring systems.
- determine what production and financial record-keeping systems might work best in an expanding dairy.

Effective record-keeping systems are essential for the successful management of a modern dairy operation. Systems are needed to monitor and support the daily management of the dairy, evaluate the efficiency of the operation, and support external reporting needs. Keeping records is not the objective; it is the analysis and use of the information contained in them that is important.

A key role of the manager is to plan strategically and use resources in a way that leads to a profitable and sustainable dairy enterprise. To do this, the dairy manager must thoroughly evaluate all possible strategies and their impacts. Good records allow the manager to make both daily and long-term decisions. Dairy producers over the years have had a large selection of production and financial systems to choose from. Historically, records were manually kept and analyzed, but in recent years electronic equipment has enabled the manager to automatically monitor many of the processes in the dairy.

Selection of record-keeping systems should be an integral part of the planning process since it will impact the labor needs and equipment costs. The number of systems selected should be based on the operator's objectives, plus the cost, use requirements, and benefits of each system. Remember to consider and budget for the time needed to install and maintain such systems. Train users of new systems on their use *before* finishing the modernization, and, if possible, have existing databases converted to the new systems before moving into updated facilities. Learning to use (or training someone else to use) a new system takes time and concentration and should be accomplished before the start-up process.

Uses of Data for Decision Making

Planning and control of operations requires information that describes past performance, monitors ongoing performance, and forecasts future performance, allowing the manager to select and implement appropriate actions in the continual process of adjusting while seeking to achieve business goals. Both strategic (long-term) and tactical (day-to-day) decisions must continually be made. Often farm records are used to produce descriptive (summary) information about the performance of the operation. The usefulness of this descriptive information depends on whether the data summarized are timely and under the manager's control. When this descriptive information is combined with norms and standards from external sources (comparative analysis with other dairies) or internally generated management goals, diagnostic information can be generated. Historically, a disproportionate effort was made in the production record-keeping area, but recently several financial benchmark databases have been created and are being used extensively by producers to identify management areas to emphasize.

In addition to descriptive and diagnostic information, records provide predictive information for planning, and prescriptive information (advice) for improvement. The data gathered and summarized by a good record-keeping system can be used as input in computer programs that are designed to help dairy managers make sound management decisions (decision aids).

Types of Record-Keeping Systems Maintained by Dairy Producers

Dairy producers have a wide array of both production and financial record-keeping systems to select from. Production record choices include dairy herd improvement (DHI) records maintained through **dairy record processing centers (DRPCs);** on-farm **dairy herd management (DHM) systems;** and **electronic dairy data collection (EDDC) systems,** which support automatic recording of milk yield, animal activity, milk quality, and so on. Financial records are often maintained manually, using on-farm computer systems, or are sent off-farm for remote entry and analysis. The value of these records lies in when and how the producer uses the information they contain.

The systems that are currently being used or developed monitor animal production, status, or behavior; equipment usage and status; or employee effectiveness. Computer systems combining the information from these milking-, feeding-, and labor-reporting systems have the potential to greatly increase the efficiency of a dairy operation. The following are some of the choices producers should consider and budget for during the modernization planning process.

DHI Records

The DHI system is one of the oldest and most widely used databases in the dairy industry. Data collected on individual farms are not only used by the herd manager to make on-farm management decisions, but are shared with the industry. Artificial breeding organizations (AI), breed registry organizations, and research groups use these data for cow and sire evaluations and research efforts. This sharing of information continues to be a unique feature that adds a huge value to the industry.

DHI is an excellent tool for monitoring a herd's performance. It maintains a complete history of each animal's identification, production, and udder health, plus health and reproductive information. One reason DHI has been successful for so long is that its organization supplies a field person to collect the data on a regular basis. The rolling herd average (RHA) information that is generated standardizes production performance on a per-cow basis. The RHA milk production allows the user to monitor progress over time but has limited value in predicting herd profitability. The RHA number-of-cows value is very important in that it reflects the true average number of cows each period and is therefore beneficial when evaluating the financial or production performance of the herd.

Since data summarized by DHI organizations are collected periodically, they have limited value when you are making some individual animal management decisions. To avoid this problem, producers often use on-farm computers and DHM software to maintain current on-farm databases. Information in the two databases can be electronically exchanged to support both on-farm and industry needs. DHI offers a somatic cell count (SCC) program that helps producers monitor the mastitis level in their herd and implement mastitis-control programs based on individual cow information. This is the only source of such information to dairy producers and is a major reason for membership. Another good reason for participating in the DHI program is the discount on AI bull semen given when daughters are properly identified and their records are used in the national sire proving system.

On-Farm Dairy Herd Management Software

There are several DHM software programs on the market (DairyComp 305, DairyQuest, etc.). Selecting a program that is widely used in your area is important, in that local veterinarians, breeders, and other support people often will be able to access your files and help make pertinent management decisions without any training. Select a DHM program that has been thoroughly tested, to minimize computer-related problems. These stand-alone systems contain extensive information about each animal's identification, reproduction, health, and milk production. Normally, information can be entered directly into the computer or retrieved from other remote computers (Figure 16-1). Two-way exchange of information between on-farm computers using DHM programs and milking parlors, DHI organizations, bull studs, and so on, is common.

These systems normally are used to summarize information and to generate management-by-exception and herd summary reports. Since the

FIGURE 16-1 Timely updating of animal files is important. This large herd has a computer station at each animal workstation, and events are entered immediately to ensure that animal records are accurate.

information is entered on-farm, each animal's record should reflect its current status. With this type of system, procedures should be developed and implemented to ensure that the database is updated frequently, so people using the records have current information to work with. Frequent database backups should be made to avoid the potential loss of information with a computer malfunction. It is important to define the events that should be summarized to support your management style, since once event information is lost it may be impossible to go back and recreate some values. Select a system that allows you to define the values that are summarized and retained. For example, you may want to summarize at each month end the average milk production for different groups of cows and monitor the performance of these different groups over time. Since most systems by default only retain an animal's current location, the ability to recreate these values would be lost if the group summary were not defined, because you would not be able to determine which animals were in those groups at previous times.

Electronic Dairy-Data Collection Systems

Milking-parlor electronic identification and milk metering systems have been marketed for several years as tools to improve animal management through improved heat detection, earlier disease detection, and more accurate production monitoring. Operators investing in this technology hope to use daily milk weight information to refine ration-balancing, pen-change, and culling decisions. They also expect to use the variation in milk weights to identify animals that are potentially sick or in estrus, for inclusion on management-by-exception lists. It is felt that early identification and intervention into problems will decrease their severity and effect. With the high production levels expected in modern dairies, knowing the production level of individual groups of animals helps to monitor each group and identify potential problems. Some herds that do not have individual-cow electronic identification use **milk-flow meter** recording systems (Parlor Watch 305, etc.) to monitor group production averages. With this type of system, milkers must indicate when group changes occur and may need to manually identify the number of animals milked in each group. This system requires a lower initial capital investment than individual cow identification systems and can provide average milking time, temperature, and milk production information.

Although this equipment is normally sold for milk weight monitoring, it can supply additional information of benefit to the herd manager. **Unit attachment time,** duration of **unit on-time, milk flow rates,** time of **peak milk,** and parlor stall occupancy all can be used to monitor and manage animals, equipment, and employees. Being able to identify such management problems as animals in the wrong pen, milk stalls not functioning properly, or workers not reattaching units can be extremely beneficial to the dairy manager.

Consider the accuracy of identification and the total system cost when deciding to purchase an electronic identification system. The failure to read one animal's identification number is tolerable, but only if it does not start a chain reaction that results in several animals having incorrect information stored in their records. With an electronic identification system, a sort gate that sorts preselected animals to a catch pen can be a valuable secondary use for the identification. Even in herds that do routine treatment activities with self-locking manger stalls (breeding, vet checks, bST injections, etc.), this feature can be beneficial in large herds to select animals to be culled or moved. Some EDDC systems incorporate pedometers as activity monitors that measure the number of steps walked per day by an animal. Changes in steps per milking interval are helpful in both heat and disease detection. These systems are marketed by several milk-machine companies and can be purchased with or without milk-metering equipment. Research and experience indicate that these devices can be effective in identifying lame cows, cows off-feed, and cows in estrus.

Milk conductivity measurement equipment is sometimes incorporated in EDDC systems to help identify animals with mastitis. It is known that udder inflammation (mastitis) is often accompanied by an increase in the concentrations of sodium and chloride, and hence a change in the electrical conductivity of milk. Sensors for measuring milk conductivity for each quarter have been integrated into robotic milkers. Milk-temperature sensors have also been incorporated in some robotic milking systems to aid in the detection of illness or estrus.

Automated systems that monitor milking-system parameters have proven to effectively monitor pulsator function, **vacuum stability,** and bulk tank milk temperature. Such a system could continuously monitor all pulsators in a milking parlor, warn the user when any one is malfunctioning, and allow the system to be serviced before it affects cow health or milk production. Logging the temperature of the bulk-tank milk could help identify malfunctioning cooling equipment and prevent losses due to spoilage.

Feed-Monitoring Systems

Computerized concentrate feeding systems were one of the first computerized animal-activity monitoring and feed-control systems to be commercially available. They were very popular in the United States before the concept of total mixed rations (TMR) became popular. These systems are still used in many European countries, where cows are housed in freestall facilities, and small herd size makes grouping of animals by production level prohibitive. These systems are also used in conjunction with robotic milking systems to encourage cows to come to the milk stall.

Several herd feed-mixing and feed-monitoring systems are currently on the market (E-Z Feed, Feed Watch, TMR Tracker, Feed Supervisor, etc.). They are used in dairies where the TMR delivery vehicle is equipped with a computerized scale. This system prompts the feeder as to the order and the amount of each ingredient to be loaded for each ration, and monitors feed loading and feed unloading. It generates loading and unloading errors by day, operator, feed type, and pen, and allows entry of feed purchases and sales, used in conjunction with feed loading weights to maintain an ongoing feed inventory for the dairy.

The Internet

The Internet is a system that connects networks of computers across the world and includes both the World Wide Web and electronic mail (e-mail), which allows messages to be sent instantly and at little or no charge, to any other user anywhere in the world. Use of e-mail to communicate and exchange information can greatly increase your efficiency if you own a computer and subscribe to an on-line service provider. Accessing the World Wide Web can be a very effective way to find information and purchase materials. According to an Agriculture Department National Agricultural Statistics Service (NASS) report, 43 percent of U.S. farms now have Internet access, compared to 29 percent in 1999 (USDA, 2002). Other data from the report showed that nearly 55 percent of farms had access to a computer in 2001, compared to the 1999 level of 47 percent; 50 percent of all U.S. farms own or lease a computer, up from 40 percent in 1999; and farms using computers for their farm business increased from 24 percent in 1999 to 29 percent in 2001.

Other Animal Production Systems

Several large dairies have installed water meters that can be accessed electronically to record the water consumption of individual groups. Temperature and humidity values can be read and recorded for both inside and outside of the animal housing units. Automatic body-weight measuring systems, which determine the weight of each animal as it passes through an access lane, have been installed. Decisions based on information from these sources could help identify malfunctioning equipment, management changes needed, or animal health problems.

It is obvious that electronic herd-monitoring systems supply data that contain valuable information pertaining to individual animals; those needing attention can be identified and added to management-by-exception lists. These data, if correctly interpreted, can also aid in the detection of group-wide or herd-wide problems. One problem experienced by some users is that the normal variation in these types of data is large, which complicates their use in dairy herd management. For example, milk weights vary from day to day, but a large deviation from expected milk weights may not always signal the onset of estrus or mastitis. Computer software normally provides initial threshold levels for selection of animals to be added to management-by-exception lists. Cows on these lists should be monitored and selection criteria modified to reflect observed and desired accuracy.

Financial Record-Keeping Systems

Modern dairies must maintain detailed financial records to monitor current status and progress. Both on-farm programs and monthly mail-in services exist; the cost, security of data, and timeliness of results should determine the choice of a system. The key to any system is to have a **chart-of-accounts** that specifies business areas you want to analyze. When you select the chart-of-accounts for your business, follow the main headings used by benchmark databases from your local area. Financial software designed for general business use tends to be less expensive because of the large number of users in its target audience; financial software written specifically for agriculture may be more expensive because of the limited market. General programs (QuickBooks, etc.) allow you to define your own chart-of-accounts that matches your needs, and are easy to learn and use. The system must allow you to enter both the dollar amount and the number of units associated with a transaction—for example, amount of milk sold (cwt) and money received ($). This will allow you to calculate and monitor cost per unit of production.

Computerized financial records allow you to easily update financial reports and monitor the financial health of your operation. If your operation consists of multiple enterprises (cows, heifers, crops, etc.), maintain records that support the analysis of each enterprise. If the income statements generated for your operation are to be accurate, inventory adjustments must be made. Establish a procedure to do a monthly feed inventory, as this information increases the accuracy of your income statement and allows you to modify rations to meet inventory constraints.

Benchmark Databases

Many organizations have developed benchmark databases that can be used to diagnose the financial performance of dairy operations and identify areas needing improvement. Gary Frank (1999a) analyzed the 1998 records of 780 dairy farms in the state of Wisconsin in one such database. When these data were analyzed by herd size, the average production and total margin per farm increased as herd size increased. Smaller herds averaged 39 cows, produced 17,096 pounds of milk, and had an average return on assets of 4.2 percent. Medium and large herds averaged 69 and 251 cows, produced 19,586 and 21,180 pounds of milk, and had an average return on assets of 7.9 percent and 11.6 percent, respectively. Group analysis of physical parameters (herd size, milk production levels, etc.) and financial performance values (net returns per cow, return on assets, etc.) were poorly correlated. This lack of correlation may be caused by many different factors, which could include (1) the different goals of producers, (2) different farm management systems involved, (3) the fact that financial records are often maintained for tax reporting purposes and not management purposes, and (4) the fact that the records do not reflect changes in the value of assets resulting from different spending levels on maintenance and improvements. These findings lead one to believe that benchmark database values may be beneficial for evaluating the performance of individual dairies but that little accuracy can be expected when attempting to predict the profitability of an individual operation based on its size or production level.

Limitations of Tying Costs to Events

Most producers would like to know, and could benefit by knowing, the exact costs of different activities. Tactical decisions could be improved if this type of knowledge was available; for example: Should an animal, identified with a specific health problem, be treated? The correct management decision would depend on treatment costs, expected benefits, and the probability of recovery. The cost of treatment could include the extra direct costs (veterinarian farm-visit charge, medications, labor) and indirect costs (loss of income from contaminated milk or carcass). The benefits could include additional productive time in the herd. The probability of recovery could be estimated from previous experiences on this or other farms. Good records could provide the information needed to make such a decision, but keeping records with sufficient detail is actually difficult because the time it takes to collect and record the information is worth more than the perceived value of the information.

Some of the difficulties of trying to collect detailed information are suggested by the following questions.

- How can accurate time data regarding workers and equipment performing specific tasks be recorded and stored?
- How should costs that are commingled (for example, a veterinarian bill for treating and vaccinating several animals on one farm visit) be allocated?
- How should unaccounted worker, equipment, or facility costs be allocated?

The answers to such questions as these must be determined by each producer, based on knowledge of the operation and of how assets are deployed. The expense to be allocated is often based on the relative income generated by the specific enterprise. The major portion of the expense may be allocated to the primary enterprise, and the remainder divided between other enterprises that make use of the asset. To define time-use values, managers will often define normal usage patterns and then monitor activities for short periods to establish realistic time-use percentages for allocation of costs to enterprises.

Information Types

The following types of information can be collected by current record-keeping systems and may be considered when planning a dairy modernization project.

Herd inventories—current inventories of milking cows, dry cows, and heifers by location

Herd production—actual milk shipped per day

Herd milk level—bulk-tank averages by pen, parity level, and so on

Herd components—somatic cell count, butterfat percentage, protein percentage, and so on

Herd culling history—history of level and reason for culling

Cow milk production—current and historical milk and component levels

Cow pedometer readings—current and historical activity measurements

Cow milk composition—current temperature, conductivity, and component information

Cow health—current status, treatment history, and effects relating to previous treatments

Cow reproduction—current status, breeding and treatment history, and effects relating to previous events

Animal movement information—cows sold, died, turned dry, moved, and so on

Feed inventory—purchases and sales, usage updates, weigh-backs, plus the ability to periodically calculate and rectify feed inventory shrink values

Feeding programs—rations and mixing recommendations by pen

Feed-mixing program feedback—loading and unloading information

Labor use and compensation level—details on hours worked and hours per activity performed

Financial status—loan balances, accounts receivable, accounts payable, and projected income and expense streams

Financial details—all purchases, sales, payments made, and payments received

Marketing information—current and future projected values of milk, calves, cull cows, replacements, feeds, and so on

Equipment monitoring—bulk-tank temperature, pulsation function, and vacuum-stability data

Environmental information—water-consumption level by pen, barn-temperature information, and so on

Manure historical information—time frames, animal units involved, and amounts hauled

Examples

The following list contains examples of the types of warnings that can be generated by record-keeping systems to aid in the management of a large dairy. Monitoring of activities and results can identify employee problems, malfunctioning equipment, ineffective procedures, short- and long-term financial concerns, and so on. If both production and financial information are available, it is possible not only to identify situations needing attention but also to estimate the economic consequences of changes.

- Dry cows were not moved yesterday as scheduled.
- Pens 4 and 5 milked in different order than normal on 3rd shift.
- Bulk tank temperature over 45° from 00:12 to 04:13.
- Milking time for 3rd shift 50 minutes less than normal, and cows down 11%.
- Production/cow in pen 6 has dropped 9% the last 5 days.
- No pedometer readings for cows in pen 3, fresh less than 40 days.
- Cows bred to bull AI555 have conception rate of 3%.
- Breeder ABC has recent conception rate of 14%.
- 80% of the cows receiving treatment XYZ were sold within 21 days.
- Milk production for cows moved from pen 1 to 2 dropped 18% (77 to 63).
- Feed mixing loading deviation over 3% for 1st shift.
- $12,000 bills need to be paid today, checkbook balance $10,000.

As dairies get larger, tasks are standardized, and workers are expected to follow written protocols. In this type of environment, workers may be able to efficiently collect information relating to their jobs that could provide feedback, which is difficult to do in smaller operations.

Using the Data You Collect

It is obvious that the chief purpose of the information from record-keeping systems is to support the manager's decision-making process; it can also be used to help motivate employees. Letting the employees know the goals (Figure 16-2)

FIGURE 16-2 Letting your employees know what is expected of them and giving them credit for a job well done leads to employee satisfaction and enhanced performance.

and current performance of the operation (Figures 16-3 and 16-4) can instill pride. Financial information often cannot be shared, but sharing production-related information may improve communications and work performance.

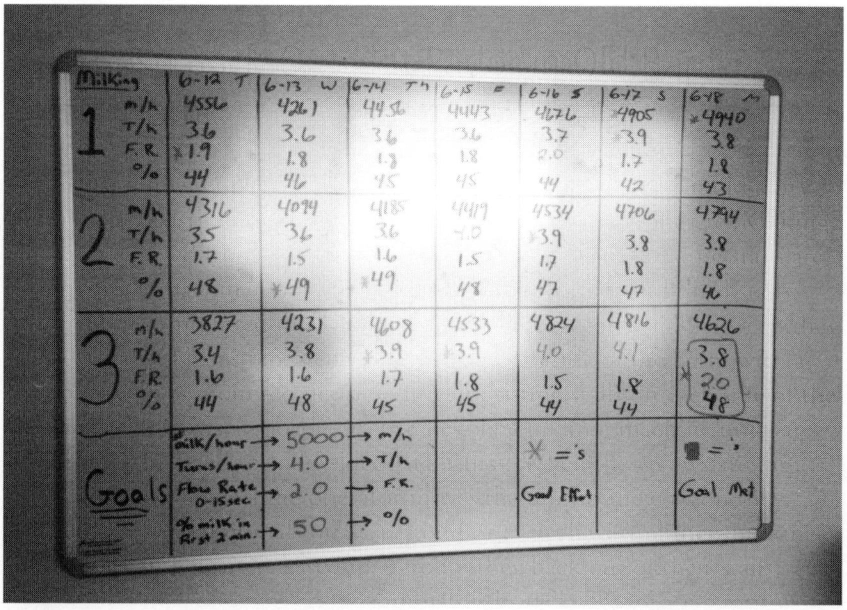

FIGURE 16-3 Bulletin boards and wall charts can effectively summarize information and distribute it to employees.

FIGURE 16-4 Monitoring results and posting them for employees to see can be an effective motivator.

SUMMARY

When an existing dairy is planning to modernize its operation or a new dairy is being planned, the labor and tools needed to monitor the operation should be determined and resources allocated to ensure their success. The type of record keeping selected must allow the manager to constantly monitor the status of the operation and make decisions that optimize its operating efficiency and profitability. When data-collection and -recording systems are being selected, the initial cost, labor requirement, and anticipated use of the information should be considered. Purchasing decisions can be very difficult because the cost often is much easier to determine than the expected benefits. Prioritize decisions based on expected returns, with the goal of optimizing the use of resources. Calculating the expected change in profitability or cost of production requires production and financial information. Remember to plan for training and the creation of the initial database, which is often a slow process; if possible, create the initial database before cows are added.

CHAPTER REVIEW

1. When is the ideal time to train users of a new record-keeping system? When is the ideal time to convert existing databases?
2. What two types of information are required to make strategic and tactical decisions and to produce descriptive and diagnostic information?
3. List the four types of information that the DHI system collects for every animal it monitors.
4. How might a dairy operation fail in using its electronic record-keeping systems? List the oversights and errors that could create inaccurate records.
5. What are the three primary health issues that EDDC systems help to detect?
6. List the 10 most important pieces of information supplied by record-keeping systems. Why do you feel each item is important?
7. What difficulties can a farmer experience in trying to correlate costs to events?

REFERENCES

DeLorenzo, M., & Thomas, C. (1995, April). *Economic decision support systems for dairies.* Presented at the Second Western Large Dairy Herd Management Conference, Las Vegas, NV. Las Cruces: New Mexico State University Cooperative Extension Service.

De Vries, A., & Conlin, B. J. (1997, September). *Dairy herd management strategies for improved decision making and profitability.* [Handout]. North Central Cooperative Research Project—NC119. St. Paul: University of Minnesota, Department of Animal Science.

Eicker, S. W., Fetrow, J., & Stewart, S. C. *Marginal thinking: Making money on a dairy farm.* [Handout]. Ithaca, NY: Author.

Eicker, S. W., & Stewart, S. C. (1998). Computerized parlor data collection and use: Monitoring the cows, the people, and the parlor. Verona, WI: National Mastitis Council.

Fountain, F. M. (1999, April). Developing the optimum financial structure for your dairy operation, or, Improving profitability and survivability through financial management: A lender's perspective. *Fourth Western Dairy Management Conference,* Las Vegas, NV. Manhattan: Kansas State University, Department of Animal Science and Industry.

Frank, G., & Vanderlin, J. (1999a). *1998 financial benchmarks on selected Wisconsin dairy farms.* [Handout]. Madison: University of Wisconsin Cooperative Extension.

Frank, G., & Vanderlin, J. (1999b). *Milk production costs in 1998 on selected Wisconsin dairy farms.* [Handout]. Madison: University of Wisconsin Cooperative Extension.

Palmer, R. W. (1999, April 22–26). Management systems to improve the production and financial performance of dairy farms. *Fiftieth annual meeting of the European Association of Animal Production,* Zurich, Switzerland.

Rossing, P. H., Hogewerf, P. H., Ipema, A. H., Ketelaar-De Lauwere, C. C., & De Koning, C. J. A. M. (1997). Robotic milking in dairy farming. *Netherlands Journal of Agricultural Science, 45,* 15–31.

Soares, M., Kuntz, R., & Avila, W. *EZFEED—The dairyman's ultimate feed management tool.* [Promotional brochure]. Tulare, CA: Valley Agricultural Software.

Stowell, R. R., Bickert, W. G., & Nurnberger, F. V. (1998, January). *Radiant heating and thermal environment of metal-roofed dairy barns.* Presented at the Fourth International Dairy Housing Conference. St. Joseph, MI: American Society of Agricultural Engineers.

Sumrall, D. P. (1999, April). *Management strategies for dairy systems.* Presented at the Fourth Western Dairy Management Conference. Las Vegas, NV. Manhattan: Kansas State University, Department of Animal Science and Industry.

U.S. Department of Agriculture. (2002). *Census of Agriculture.* www.nass.usda.gov/census/census/. USDA-NASS, Room 5829-South, Washington, DC 20250.

Chapter 17

Acquisition of Products and Services

KEY TERMS

asset
bare cropland
barn cleaner
bottom-line profitability
buy-sell contract
contingency plan
contracted forage
cover forage
custom cropper
custom milker
custom operator
dry matter (DM) basis
farming out
fermentation shrink
late-payment penalty
lease rate
loaded mile
long-term contract
nutrient build-up
partnership arrangement
relative feed value (RFV)
right of refusal
step-by-step approach
supplier

Objectives

After completing the study of this chapter, you should be able to

- identify the elements of a dairy operation that can be contracted out.
- identify reasonable prices for equipment and building rental.
- identify the advantages and disadvantages of contracting for feed and animal replacement.
- incorporate the idea of contracting for services into an existing dairy's modernization plan.

A successful dairy requires that the operator assemble the **assets** (land, building, equipment, cattle, feed, and labor) needed to run the dairy. For many producers, modernization means a complete change in how the business operates: freestall barns, milking parlors, flat feed storage, and automated manure-handling systems replace stall barns with pipeline milking and **barn cleaners.** This abrupt change in system components, along with the dramatic increase in herd size needed to justify the investment, often requires the producer to look at alternative ways to acquire the assets needed. This has led dairy producers to several basic philosophy changes: from thinking about owning assets to controlling assets, from needing production skills to business skills, from selling on the open market to contracting, from working as an independent operator to partnering, and from diversified operations to specialized operations.

As you plan your modernization, you must consider more than just the facilities needed. During the planning process, a decision must be made relating to owning versus leasing, growing crops versus buying feed, planting and harvesting crops versus having a **custom operator** do it, raising heifers versus having a custom heifer raiser raise them, equipment sharing, and so on. Contracting for services often allows a producer to specialize in an area in which he or she is most qualified, cut labor requirements, and reduce investments. When planning a major change to a business, base your decisions on the investment required and its associated risk. Possible enhancements should be prioritized and selected based on their expected returns and their value in helping to achieve long-term goals.

Partnership Arrangements

Partnership arrangements, in which people join forces and share resources, are not new to the dairy industry. There are many ways to structure and organize a partnership, based on the objectives of the individuals involved. The most common way is for two or more people to pool their resources and divide income according to each partner's inputs. This allows producers with limited resources to be part of a larger operation and thereby take advantage of the increased scale of the operation created by the partnership. An example of a successful partnership is a situation in which a crop-producing partner supplies the forage and dairy facilities while the dairy partner supplies the cows, labor, and management, and the two split the milk check. Many family partnerships are based on the older generation supplying the assets and the younger supplying the labor and management. Whatever arrangement is proposed, it behooves everyone involved to treat all participants equitably if the arrangement is to last.

Enterprise Choices

Most dairy operations traditionally included three enterprises: milking cows, heifer raising, and crop raising. In addition, each enterprise performed many different functions: milking, manure hauling, animal estrus detection and breeding, hoof trimming, heifer raising, crop planting and harvesting, and so on. As herd sizes have increased, producers have chosen to specialize in enterprises that most interest them. This specialization allows producers to maximize the use of borrowing capacity and management expertise. For some, this means increasing the milking herd, having heifers custom raised, and hiring **custom croppers** to plant or harvest all or part of their crops. For others, it implies selling the milking herd and becoming custom heifer raisers or custom croppers. Enterprise accounting information from your financial record-keeping system should help you define the relative profitability of different enterprises and can help you determine which should be continued.

As operations increase in size, producers often rely more on outside **suppliers** to perform many functions. These outside service providers can specialize in specific functions, more fully utilize assets, and perform services at a lower average cost than the producers can themselves. Hiring custom operators to do specific functions minimizes employee training and avoids disrupting employees from their routine tasks. The availability of custom operators and associated costs must be investigated. Since resources are often limited, **farming out** some jobs can often cut capital needs and labor requirements, leading to increased **bottom-line profitability.** Remember that when custom-service providers are hired to perform specific functions, you must evaluate not only the cost associated with the decision but the likelihood that providers will perform the activity on a timely basis and in a satisfactory fashion. Insist on written agreements that specifically define the terms of service, and **contingency plans,** in case things do not workout as planned.

Types of Custom Operators

Custom operators are available to support cropping, heifer raising, and herd management functions. Feeding the dairy herd is the most expensive single line item for a dairy. Finding the best solution to supplying the feed depends on the amount of land currently controlled (owned and rented), additional land available (which can be purchased or rented, or from which feed can be gotten), herd size, and the expected delivered cost of purchased feeds. The amount of land needed, if crops will be grown locally, depends on geography and local weather conditions. Manure disposal is critical and requires sufficient land to avoid **nutrient build-up.** The proximity of the dairy herd to land for crop growing and manure disposal must be considered if wet forage is to be transported to and wet manure transported from the dairy.

Many of the herd management functions, which historically have been done by the herd owner or herdsperson, are now being performed on a contractual basis. AI breeding organizations will check the herd daily and identify and breed animals according to your breeding program. Hoof trimmers will establish schedules and trim animal hooves on a regular basis. Some veterinarians will work on a retainer basis. **Custom milkers** have been hired by some producers and are paid according to the amount and quality of milk harvested. Part of the planning process should include investigating the services available in your area and determining how they fit with your management style and objectives. Use of support services can reduce labor requirements and increase the effectiveness of herd reproduction and health programs. Understanding the needs of contract service providers may influence facility feature selection. A good example is the installation of self-locking manger stalls, which support arrival-time flexibility, allow several different functions (breeding, vet work, etc.) to be performed at the same time, and can result in less on-farm time for suppliers.

Estimating Agricultural Field Machinery Costs

During the planning process you must learn the cost, or potential cost, to own and operate specific farm machinery. This information is needed to determine if machinery should be purchased, and can be helpful when negotiating contracts with custom operators. The most accurate method of determining machinery cost is the record of the actual costs of previously owned equipment. If this information is not available, machinery-cost estimates can be found at the University of Wisconsin Extension Web site: http://www.uwex.edu/ces/ag/. Remember to consider both fixed costs (those costs that are incurred regardless of use) and variable costs (those that vary with usage). Include the expected depreciation, interest, insurance, housing, and tax costs associated with each piece of equipment.

Establishing Building Rental Costs

Often producers can rent complete dairy facilities or individual buildings to house all or part of the operation's cattle. Used dairy facilities usually rent for $10–18 per cow per month, and new facilities may be $25 or more per cow per month. Buildings for heifers range from $2–6 per head per month, depending on animal size. To arrive at an equitable **lease rate,** the value of each asset, current interest, tax and insurance rates, plus expected repair costs should be considered. Repair costs of new buildings normally are about 3.5 percent of initial cost; figure 5 percent for older buildings. Taxes of 1 percent on farm buildings, and insurance at 0.5 percent of initial cost, are common. Table 17-1 contains an example of a used 300-cow facility valued at $1,500 per stall; a double-8 parlor with equipment valued at $5,000 per stall and building valued at $5,000 per stall; current interest rate of 9 percent; no property tax on equipment, but 1 percent on buildings; and insurance at 0.5 percent of original value. The value of the freestall barn includes improvements such as a well, manure storage, and feed storage (these values may not represent current values in your area). Note that since assets are depreciated to zero value over their lifetime, the interest rate used is one half of the current rate (i.e., interest payments would decrease each year and on the average be one-half of the starting value).

TABLE 17-1 Example lease payment calculation for a 300-stall facility with a double-8 parlor.

Asset Type	Freestall Barn	Parlor Building	Parlor Equipment	Total ($)
Value/stall	$1,500	$5,000	$5,000	
Number of stalls	300	16	16	
Current value	$450,000	$80,000	$80,000	$610,000
Current value/cow @ 110% stocking				$1,848
Years of life	15	15	5	
Depreciation	6.7%	6.7%	20%	
Interest*	4.5%	4.5%	4.5%	
Repairs	3.5%	3.5%	3.5%	
Taxes	1%	1%	0%	
Insurance	0.5%	0.5%	0.5%	
Lease rate	16.2%	16.2%	28.5%	17.8%
Lease ($/year)	$72,750	$12,933	$22,800	$108,483
Lease ($/cow/yr) @ 110% stocking				$329
Lease ($/cow/mo)				$27.39
Lease without repairs ($/cow/mo)				$22.00

*Interest rate shown is one-half the current rate, an average reflecting depreciation over the term of the loan.

Calculating Feed Requirements

When planning the feed needs of a proposed dairy, consider the grouping strategies and feeding program to be followed. Estimate the number of animals and their intake requirements to support maintenance, growth, and production. Calculate the different losses involved in forage harvesting, storage, handling, and animal feed refusal. This step is often overlooked in planning, and many herds have been hurt the first year after expansion due to insufficient forage or substandard quality.

Work with your nutritionist or county agent to establish a sound feeding program and realistic inventory needs. The proper level of forage ingredients should be determined. For example, feeding trials generally demonstrate similar milk production from cows fed diets based on either corn or alfalfa silage, but the use of a single forage source may require better herd management or at least different operating procedures to prevent adverse affects on cow health or milk production. Many people feel that the use of at least one-third of each forage type helps reduce the risk of crop loss and sick animals.

TABLE 17-2 Yearly forage needs for a 600-cow herd.

	Tons (as fed)	Tons (DM basis)	Expected Yield/Acre	Acres Needed
Hay	1,750	1,488	4	372
Alfalfa haylage	5,665	2,549	4	637
Corn silage	7,399	2,590	6	432
Total		6,627		1,441

For high-producing cows, a 5–10 percent feed refusal rate is a suggested estimate. Consider how this refused feed will be used. Caution must be exercised if it is fed to heifers or dry cows because of its nutrient level and variability. Table 17-2 is an example of forage requirements calculated for a 600-cow herd feeding a 50:50 corn silage and alfalfa haylage diet, 504 milk cows (4 rations), 96 dry cows (2 rations) and 550 heifers (3 rations), with a 15 percent harvest loss. Using the expected forage needs, average crop production levels, and losses expected during storage and feeding—plus some allowance for annual crop variations—an estimate of the amount of feed or acreage needed to produce it can be generated.

Buy-Sell Agreements for Crops or TMR

Producers who expand their operations often can reduce their capital investment and ensure sufficient feed through **long-term contracts** with local producers to supply part or all of their forage needs (known as **contracted forage**). Some producers even buy a complete TMR that is delivered to their operation daily (Figure 17-1). These **buy-sell contracts** often include provisions for the manure disposal on the land used to grow the contracted crops. The value of dairy manure can be determined based on its nutrient value and removal costs.

When buy-sell contracts are being developed, it is important to establish an equitable pricing system that is less variable than the open market. The price of forages should never be as high or as low as the open market extremes but should result in approximately the same average price. Price may be based on historical crop values or cost of production, or it may be tied to market prices of other commodities such as corn grain and soybean meal. Specify the amount and quality of each feed desired and include penalties if either of these criteria is not met. Prices should be based on the amount and quality of forage dry matter delivered.

Table 17-3 shows an example calculation of the value of corn silage in the field. Starting with the assumptions that the corn is worth $2.50 per bushel and the expected land yield is 140 bushels per acre, $15.30 per ton would be considered a fair price for the corn silage given the harvesting,

FIGURE 17-1 Buying a complete TMR can allow a herd owner to specialize in milking cows without all the costs and risks associated with raising crops.

TABLE 17-3 Value of corn silage sold in the field.

Base price	$2.50/bu
− Harvesting	$0.15
− Drying	$0.16
− Storage and handling	$0.15
Net price	$2.04/bu
Value @ 140 bu/acre	$285.60/acre
+ Extra seed cost	$8.70/acre
+ Extra potassium	$12.60/acre
Cash return needed	$307.00
Value/ton @ 20 ton/acre	$15.30/ton

storage, and handling values shown. Notice how the base price is first adjusted down by considering the money saved in selling corn silage rather than corn grain (avoiding harvesting, drying, storage, and handling costs) to arrive at a net price, then any extra costs associated with corn silage are added to get to a realistic cash return per acre.

The most important consideration when establishing any contract is that a trusting relationship is formed and that each party enters the relationship in good faith. A contract should always be written to ensure that there is no misunderstanding by either party, and checked by a lawyer to

TABLE 17-4 Actual cost of 7331 tons of corn silage purchased by a large dairy.

	Total Cost	Cost/ton without Shrink	Cost/ton with 10% Shrink
Initial cost ($17.50/ton)	$128,293	$17.50	$19.44
Custom chop ($185/hr + fuel)	$18,800	$2.56	$2.85
Trucking	$9,250	$1.26	$1.40
Packing ($120/hr)	$9,495	$1.30	$1.44
Covering (plastic & labor)	$1,180	$0.16	$0.18
Inoculent	$8,635	$1.18	$1.31
Miscellaneous	$3,360	$0.46	$0.51
Total cost	$179,013	$24.42	$27.13

ensure that it is binding and that methods of termination are acceptable to both (or all) parties. The dairy producer must be confident in the grower's management skills and ability to produce a product of the quality desired. The grower must be assured of payment on a timely basis. Contracts should specify payment methods, interest rates, and any **late-payment penalty** rates. Often crops purchased are paid for in 13 payments, one payment at planting or harvest time and one payment per month for a year. This allows the producer to avoid a large investment in inventory and levels cash-flow requirements; it also gives the grower a constant income stream. If feed is to be purchased in the field and harvested by the dairy producer, all handling costs should be considered. Table 17-4 shows the actual cost to a large dairy that purchased corn silage in the field for $17.50 per ton. After all costs to harvest, haul, pack, and **cover forage** were considered, along with a 10 percent estimated **fermentation shrink,** the corn silage cost the producer more than $27 per ton.

Advantages and Disadvantages of Forage Contracts

With forage contracts, unlike land rental agreements, the landowner or grower can farm his or her land and earn a competitive return for labor and management. Additional yield quality and quantity can lead to increased returns. Corn silage is a logical crop to contract because the risk associated with it is relatively low compared to other crops. Corn silage yields follow closely those of grains and are less sensitive to conditions that impact plant maturity (such as a late planting date, summer weather conditions, or an early fall frost). With a corn silage contract, the harvest price is known at the beginning and is not subject to the volatility of the grain market. Since corn silage is harvested earlier in the fall than corn grain, additional time is provided for fall tillage.

For the dairy producer, in areas where land for rent or sale is limited, a forage contract may offer the only alternative for meeting forage needs on acreage that is within a reasonable proximity to storage facilities. A forage contract also enables the dairy producer to acquire additional feed without expending time, labor, and machinery during the planting or harvesting seasons, when timeliness is critical. Knowing what will be paid for forage prior to the growing season can make budgeting easier.

One disadvantage for the contracting grower is that since the price is fixed, he or she cannot take advantage of unanticipated price increases between planting and harvest. This drawback could be overcome if the contract were structured to float with local market prices. The grower must realize that if the contract is written so that the price will float with the market, there is also the risk of a price decline between planting and harvest. The major disadvantage of a buy-sell agreement is that the grower must build provisions into the agreement to guarantee that payment is made in accordance with the contract terms. In essence, the producer becomes an unsecured lender after the forage is delivered if it has not been paid for.

Two potential disadvantages of forage contracts exist for the dairy producer. First, the dairy producer loses some control over the production of feeds in terms of planting, fertilizing, and weed control. Second, renting **bare cropland** and being responsible for all production inputs will often result in cheaper feed (per ton) than is obtained through a contractual agreement.

Other Buy-Sell Contract Provisions

If a contractual agreement specifies that the grower is to be paid based on yield per acre, reasonably accurate estimates of yield are critical. To get these, some or all loads of the product can be weighed, or representative strips across a field can be measured and weighed to determine yield. If forage is purchased after harvest, each load (or representative loads) should be weighed (Figure 17-2). If payment is based on forage quality, the contract should specify how each load will be sampled, how samples will be taken, and who will pay for the analysis. Often a sample is taken from each load and placed in a container that may be sampled at the end of the day. Thorough mixing of the samples before selecting a test sample is recommended. Often two samples are taken each day so that both the buyer and seller can have the product analyzed independently.

Contracted forage price should include the cost to grow, harvest, deliver, and store the forage. Determination of which of these services will be provided by the seller must be determined, and the price adjusted to reflect these services. Current trucking rates for hay delivery are $1.30 per ton per **loaded mile** for distances of 200 or more miles. Shorter trips are charged at a higher rate because of the time to load and unload. Local forage hauling is normally charged by the hour, based on the trucks' capacity.

FIGURE 17-2 If forage is to be purchased, an on-farm scale to weigh incoming feedstuffs is highly recommended.

Where will the silage be stored? Any fermentation or storage losses must be considered. Prices per ton of fermented feeds should be 5–15 percent higher than fresh-cut forage delivered to your site.

High-producing dairy cows can utilize only good-quality forage. Hay or haylage with a **relative feed value (RFV)** lower than 125 should not be accepted for the milking herd, and lower-quality stocks should be stored separately and used for dry cows or heifers. The only way to determine RFV is to sample and test delivered forage. Since determination of RFV takes time, a visual screening technique, based on forage color, stem thickness, and percentage of leaves, must be developed. Consider placing a **right of refusal** clause in a contract, which specifies under what conditions feed can be refused. The moisture level of haylage can vary considerably. You must be sure that it is not delivered too wet in the morning or too dry later in the day. Blending loads from different sources may help prevent this problem. Silage less than 30 percent **dry matter (DM) basis** has a risk of fermentation failure, and silage more than 50 percent DM may be difficult to pack correctly. Provisions may need to be included for failure to deliver product in correct moisture range. Since haylage when delivered may have different moisture levels, payment should be made on a DM basis. As with corn silage, a method of determining DM percentage must be established.

Forage suppliers may want to contract only 60–70 percent of their normal yield, to ensure that they can fulfill their contract. Contracts should include provisions for failure to deliver or delivery of extra amounts.

Heifer-Raising Contracts

Since the cost of replacement animals is the second largest expenditure on a dairy farm after feed cost, it also should be closely evaluated during the planning process. This is especially true now that the average cost of heifers has drastically increased. Two main areas must be considered: (1) management practices that minimize culling, and (2) a heifer-growing or -acquisition approach to minimize replacement costs. What is your true cost to raise an animal? What is the quality of the heifer raised? Do you have housing that allows heifers to learn to use freestalls and self-locking manger stalls? Could the feed and labor used on heifer raising be better utilized to milk more cows? If you modernize or expand, are your heifer-rearing facilities adequate? What will you do with the poor-quality forage that has normally been fed to heifers?

In the past, when dairies expanded, economic analysis indicated that owning and raising heifers was not as profitable as selling baby calves, buying replacements, and using the limited resources to support the milking herd. With your operation, this question can be best answered using your current financial records and evaluating the availability of resources. To determine the cost of raising replacement heifers, feed, housing, and labor are the primary considerations. Labor requirements can be based on current labor use, or estimated using benchmark information supplied by local universities. One such study showed the average number of heifers per labor hour was 49, with a range of 38–68. Pre-weaning heifers were 14, and post-weaning 69 per labor hour.

Custom heifer raisers currently charge $1.55–1.65 per head per day if animals are reared from 0–23 months. Rates vary for animals of different ages, based on the feed and labor associated with their care. Contracts with heifer raisers should always include the specific terms of the agreement. Issues such as death losses, hauling, treatment costs, breeding costs, growth rate expectations, and so on should be clearly defined.

SUMMARY

Producers planning for product and service acquisitions often take a **step-by-step approach** because it requires a lower immediate investment and may be perceived to have less risk. In a step-by-step approach, first determine which assets are absolutely necessary. Have a well-defined, long-term plan in place, and implement the portion that offers the greatest return. Think about controlling assets rather than owning them. Whenever possible, use long-term buy-sell agreements to purchase feed, and outside suppliers to provide services. This can help minimize capital needs, decrease labor requirements, and allow management to concentrate on key operating functions.

CHAPTER REVIEW

1. Define "specialization" as it relates to dairy modernization.
2. What are three benefits of contracting for services?
3. Identify three activities traditionally performed by dairy employees that are now commonly hired out.
4. How much do older dairy facilities typically cost to rent, per cow per month? How much do new dairy facilities cost to rent, per cow per month?
5. List two advantages and two disadvantages of forage contracts for dairy producers.
6. Are the trends of specialization and contracting for services beneficial to the dairy industry as a whole? Explain your opinion.

REFERENCES

Karzses, J. *Effects of labor efficiency on heifer raising.* Warsaw, NY: Cornell Cooperative Extension.

Palmer, R.W. (1998, March 5–6). Forage buy-sell contracts. In *Four-State Forage Feeding and Management Conference Proceedings,* Wisconsin Dells, WI (pp. 156–167). Madison: University of Wisconsin Cooperative Extension.

Rankin, M. (1997). Contracting corn silage acres. Retrieved October 4, 2004 from University of Wisconsin Cooperative Extension Web site, http://cdp.wisc.edu/jenny/crop/contract.pdf

Schuler, R. T., & Frank, G. G. *Estimating agricultural field machinery costs.* [Extension paper A3510]. Madison: University of Wisconsin.

Stellato, J. *Contract feed production arrangements.* [Handout]. Shawno University of Wisconsin Cooperative Extension.

Chapter 18

Fitting the Pieces Together

KEY TERMS

business planning process
construction manager
contracting
conventional dairy barn
farmstead
financial feasibility
general contractor
goal setting
housing capacity
initial cost
long-term viability
management grouping
ongoing support cost
parlor capacity
project management
vaccination program

OBJECTIVES

After completing the study of this chapter, you should be able to

- summarize how to conduct a dairy modernization feasibility study.
- make phasing decisions regarding enterprise development, herd size, parlor design, and facility design.
- develop supporting plans for feeding, animal replacement, manure handling, labor management, record-keeping, and contracting.
- create an accurate timeline and budget for modernization.
- predict potential problems and develop contingency plans.

The objective of this book is to give dairy producers and their support people ideas and approaches to consider when planning a new dairy or a major change to an existing dairy. It cannot answer all your questions, but it should increase your awareness of the issues so that you can gather the additional information that you need. The **goal-setting** and **business-planning processes,** facility choices (barns, parlors, feed storage, etc.), and other support items (labor, animal replacements, record-keeping systems, etc.) have all been reviewed. At this point in the planning process, it is important to pull together all the pieces and make some decisions. Cost estimates, **financial feasibility, contracting,** and **project management** are all needed to make the modernization a reality. An example will illustrate this proposed process.

Funding

Table 3-1 in Chapter 3 showed the calculation of the expected borrowing power of the Smith Family Dairy. The operators and their wives owned and managed an 80-stall **conventional dairy barn,** with a single-story addition that had been added in 1964, and 300 acres. They realized that if they were to stay in the business long-term and pass it on to their children, they would need to consider expansion. Joe, the herd manager, was also having trouble with his knees as a result of years of milking cows. All family members were concerned about missing the opportunity to watch their children's activities as they entered high school. The family hired a private consultant, and together they defined their two-family financial position by combining their financial information (Table 18-1) and did the rough calculation shown in Chapter 3. They estimated that the family could borrow between $683,000 and 1,404,143. With this information, they began discussing realistic options that related to the family's long-term goals. Later, they approached their long-term lender and shared their thoughts with him. The lender encouraged

TABLE 18-1 The Smith families' combined balance sheet before expansion.

Land—300 acres @ $1,500	$450,000
Buildings	100,000
Equipment	100,000
Livestock—100 cows @ $1,200	120,000
—90 heifers @ $700	63,000
Investment capital—savings	50,000
Current assets	883,000
Current liabilities	100,000
Equity	$783,000

them to continue planning and felt that a $500,000–1,000,000 loan with an 8–9 percent interest rate would be possible.

This example is typical, in that the banker did not immediately guarantee a loan and coached the family to be conservative in their thinking. Obviously, bankers would investigate the circumstances and ascertain that balance sheet values were accurate and previous repayment histories supported making a loan.

Family Involvement and Business Structure

Producers often consider expansion because one or more family members want to join the operation, and additional income is needed to support the families. Thirty-four percent of the producers who responded to the 1999 Wisconsin Dairy Modernization Project (Palmer & Bewley, 2000), which surveyed producers who had expanded their operations, gave this as a major reason for expansion.

Determining the number of family members to be involved in a dairy is often difficult; age, experience, interest, financial status, and ability to get along with employees and other family members must be considered. The process begins with an open discussion among all family members, in which their goals and objectives can be heard. This discussion should lead to working with an attorney to draft a formal business structure. Formal agreements should be made, and documents developed, that define how people will be rewarded for their efforts, how ownership will be transferred, and what will happen if one or more of the participants leave the business. This is the time to correct any prior inequities and to set the stage for a long-term, equitable working relationship. Defining how ownership will be transferred is critical.

In the Smith case, this planning was relatively simple because the two families owned the operation, and it would be several years before their children would be old enough to join on a full-time basis. The families met with their attorney, updated their estate plan, structured their business as a limited liability corporation (LLC), and made provisions in case of death or divorce.

Herd Size

The long-term potential of the dairy should be established early in the planning process. No matter what herd size is currently being planned, consideration should be given to the dairy's potential long-term size to ensure that if current or future owners want to increase herd size, other structures will not obstruct that growth. Short-term thinking has often resulted in buildings, manure pits, and feed storage units being placed where they prevented or increased the cost of subsequent construction. Table 18-2 shows the expected size and housing requirements for the herd associated with the Smiths' operation. This summary shows the expected **management groupings, parlor capacity,** and **housing capacity** needs for their 600-cow dairy. Table 18-3 shows the same information, but for a single barn complex with

TABLE 18-2 Herd grouping and housing requirements for a dairy with a double-12 milking parlor and housing capacity for 576 milking cows with no special needs.

Smith Family
Final Herd Size

Herd Grouping and Housing Req's, 2 Barns, D-12P

# sites	1	
# barns	2	
Pens/barn	4	
Milk cow pens	8	
		Stocking Rate
Cows/pen	72	
Stalls/pen	68	106%
Self-locks/pen	72	100%
Milk cows in pens	576	

Main parlor type	Parallel
Number milk stalls	24
Contaminated-milk parlor stalls	0
# weeks close-up hfrs housed	6
Expected turnover rate	35%
Close-up hfrs (% of total herd)	4%

Milking Capacity

Marketable Milk Cows	% of Milking	# Cows	Exp'd Turns	Cows/Hr	Min/Group	Exp'd Time
Healthy cows	92%	576	5	120	36	4.8
Slow and lame	2%	13	4	96	8	0.1
Early fresh	4%	25	4	96	16	0.3
Total w/marketable milk	98%	614				5.2

Contaminated Milk Cows	Exp'd %	# Cows	Exp'd Turns	Cows/Hr	Min/Group	Exp'd Time
Sick (2% of milking)	2%	13	3	72	31	0.5
Just fresh-cows 4d	1.1%	8	3	72	7	0.1
Just fresh-hfrs 1d	0.3%	2	3	72	2	0.0
Total w/contaminated milk		23				0.7
			Main parlor (hr/shift):			5.9

Herd Size

		# Cows
# milking w/sick, w/o fresh		626
	% of Total	**# Animals**
Dry cows	16%	119
Total cows w/dry		745
Close-up heifers	4%	30
Total cows w/dry + hfrs	104%	775

Housing Capacity

Housing—Main Barn(s)	Capacity	Stocking%	# Animals	Housing Type
Milk cows in pens	544	106%	576	46" × 8' freestalls

Housing—Special Needs	Exp'd Head	Capacity%*	Capacity	Suggested Housing Type
Slow and lame	13	100%	13	Freestalls or bedding pack
Early fresh	25	150%	38	46" × 8' freestalls
Sick	13	150%	19	Bedding pack
Just fresh cows + hfrs	10	150%	15	Freestalls or bedding pack
Maternity-group calving (4d)	11	150%	17	Bedding pack 100 sq ft/cow
Maternity-indiv pens (0.33% milking)	2	150%	3	Individual pens
	73		104	

Housing—Dry Cow	Exp'd Head	Capacity%*	Capacity	
Close-up cows (4% of herd)	30	125%	37	48" × 8' freestalls
Close-up hfrs	30	125%	38	46" × 8' freestalls
Far-off dry (12% of herd)	89	125%	112	3-row freestall barn
	149		187	

*Capacity% = Extra capacity is added to accommodate uneven calving distributions.

Source: Palmer, R. W. (2004). Example generated using the Herd Grouping and Housing Requirements decision aid (GrpReq.xls), copyright 2001, by Roger W. Palmer. Madison, WI: University of Wisconsin–Madison.

TABLE 18-3 Herd grouping and housing requirements for a dairy with a double-8 milking parlor and housing capacity for 288 milking cows with no special needs.

Smith Family
Initial Herd Size

Herd Grouping and Housing Req's, 1 Barn, D-8P

Number sites	1	
Number barns	1	
Pens/barn	4	
Milk cow pens	4	
		Stocking Rate
Cows/pen	72	
Stalls/pen	68	106%
Self-locks/pen	72	100%
Milk cows in pens	288	

Main parlor type	Parallel
Number milk stalls	16
Contaminated-milk parlor stalls	0
# weeks close-up hfrs housed	6
Expected turnover rate	35%
Close-up hfrs (% of total herd)	4%

Milking Capacity

Marketable Milk Cows	% of Milking	# Cows	Exp'd Turns	Cows/Hr	Min/Group	Exp'd Time
Healthy cows	92%	288	5	80	54	3.6
Slow and lame	2%	6	4	64	6	0.1
Early fresh	4%	13	4	64	12	0.2
Total w/marketable milk	98%	307				3.9
Contaminated Milk Cows	**Exp'd %**	**# Cows**	**Exp'd Turns**	**Cows/Hr**	**Min/Group**	**Exp'd Time**
Sick (2% of milking)	2%	6	3	48	23	0.4
Just fresh-cows 4d	1.1%	4	3	48	5	0.1
Just fresh-hfrs 1d	0.3%	1	3	48	1	0.0
Total w/contaminated milk		11				0.5
			Main parlor (hr/shift):			4.4

Herd Size

	% of Total	# Cows
# milking w/sick, w/o fresh		313
	% of Total	**# Animals**
Dry cows	16%	60
Total cows w/dry		373
Close-up heifers	4%	15
Total cows w/dry + hfrs	104%	388

Housing Capacity

Housing—Main Barn(s)	Capacity	Stocking%	# Animals	Housing Type
Milk cows in pens	272	106%	288	46" × 8' freestalls
Housing—Special Needs	**Exp'd Head**	**Capacity%***	**Capacity**	**Suggested Housing Type**
Slow and lame	6	100%	6	Freestalls or bedding pack
Early fresh	13	150%	19	46" × 8' freestalls
Sick	6	150%	9	Bedding pack
Just fresh cows + hfrs	5	150%	8	Freestalls or bedding pack
Maternity-group calving (4d)	6	150%	8	Bedding pack 100 sq ft/cow
Maternity-indiv pens (0.33% milking)	1	150%	2	Individual pens
	37		52	
Housing—Dry Cow	**Exp'd Head**	**Capacity%***	**Capacity**	
Close-up cows (4% of herd)	15	125%	19	48" × 8' freestalls
Close-up hfrs	15	125%	19	46" × 8' freestalls
Far-off dry (12% of herd)	45	125%	56	3-row freestall barn
	75		93	

*Capacity% = Extra capacity is added to accommodate uneven calving distributions.

Source: Palmer, R. W. (2004). Example generated using the Herd Grouping and Housing Requirements decision aid (GrpReq.xls), copyright 2001, by Roger W. Palmer. Madison, WI: University of Wisconsin–Madison.

a D-8 (double-8) parlor. This is very useful because the Smiths plan to build a D-8 parlor during the first phase of their expansion, then expand the parlor to a D-12 when a second barn is built. Computerized decision aids that generate information such as that shown in these two tables are available and should be used as you plan your dairy. Most consultants have generated or have access to similar programs, which can help determine the size, scope, and feasibility of a project.

Feeding Program

A large herd requires a lot of feed; if wet forage is to be handled, the cropland producing it should be relatively close. Existing and potential cropland should be evaluated based on distance to the dairy, field sizes, surface terrain, manure application potential, yield, and length of expected availability. Forage crops should be selected based on soil fertility conditions and surface terrain. (Land with severe slopes may limit the timing or amount of manure application and the removal of corn silage.) The Smith families surveyed their neighborhood and determined that they could expect at least 1,000 acres of cropland, in addition to the 300 they owned, to be available long-term. The Smiths determined that these 1,300 acres could support the feed needs of a 600-cow milking herd with replacements. Since the land would raise good corn and alfalfa, they decided to plan for a 50:50 corn silage and alfalfa haylage feeding program.

Parlor Milking Capacity

Parlor size is a key factor in determining animal group sizes, barn pen sizes, and milking parlor efficiency. Normally, parlors are built to fully utilize the efforts of each member of the milking crew. Currently, a D-8 to D-12 parallel or herringbone parlor is considered a single-person parlor, a D-20 to D-25 a two-person parlor, and so on. A single-person parlor has the potential to support a 500–700-cow herd, and a two-person parlor can support a 1,200–1,500-cow herd.

The Smith families, after reviewing their long-term goals and local land availability, decided that a long-term herd size of 600 milking cows was appropriate. Because of their borrowing potential and their desire to start smaller and grow to this size, they decided to build a D-8 parlor and grow to 300 cows initially. The Smiths were fortunate, in that their existing facility was located on a site that both supported the use of the existing facility and allowed for long-term growth. The existing dairy barn was in good repair, and the milkhouse was large enough to support initial needs. The newest part of the existing barn had no internal posts supporting the roof, which allowed for an easy conversion to a holding pen with a crowd gate. Based on this evaluation, they decided to place a low-cost D-8P (parallel) parlor in the existing dairy barn for the first phase of their expansion but reserve space near their new freestall barn for a new D-12P parlor to be built later. Part of the existing barn's stalls were removed, and space for the parlor was created. The remainder of the barn was converted to an animal treatment facility.

Site Selection and Facility Layout

One of the most critical decisions in modernizing a dairy is selecting the correct site for any additions. Determining whether an existing dairy should be expanded or abandoned depends on the long-term growth potential of the site. Will the site support the long-term herd size desired? Remember to consider milking, housing, feed storage, manure storage, and access roads. If your existing dairy is not located on a site that allows for future expansion, then you should seriously consider building on a new site if possible.

Figure 18-1 is a rough diagram of the Smith familys' long-term facility layout, which includes two barns for housing milking cows, a milking

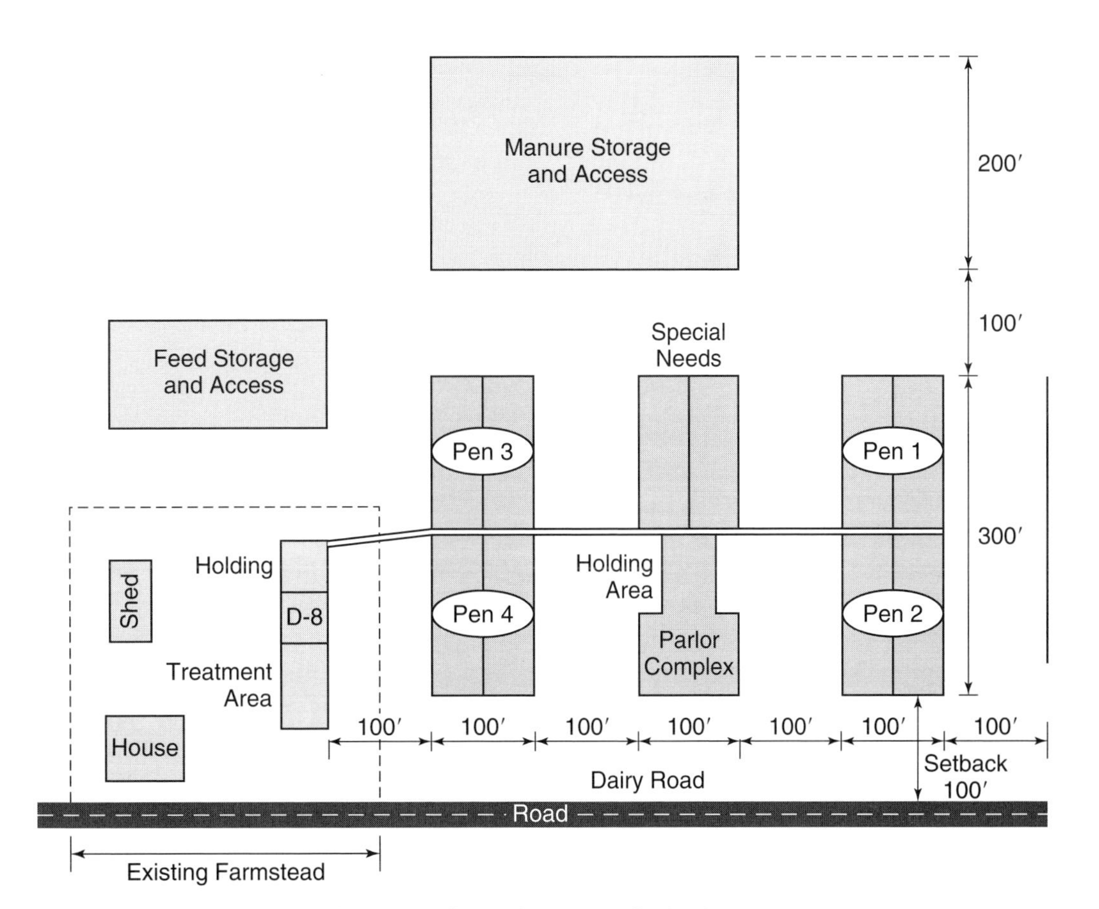

FIGURE 18-1 A diagram of the Smith familys' long-term facility layout.

parlor complex plus a special-needs barn (for cows that are dry, sick, pregnant, and so on), and an animal treatment area. Based on this rough design, the dairy is expected to cover approximately 16 acres, so they plan to reserve the complete 40-acre parcel of the existing **farmstead** for potential growth and development. Since they do not have adequate heifer housing, they plan to have heifers custom raised after the expansion and decide later if they should build heifer housing near the dairy. Dry cows will initially be housed on a second farm, but space is being reserved for future dry-cow housing as well.

Financial Feasibility

Preparing a meaningful financial analysis to evaluate the feasibility of a project requires complete and accurate information on which to base assumptions. It is important to first evaluate the long-term plan to ensure that it represents a profitable operation. Next, one must ensure that the initial phase of the plan is profitable enough to allow the operation to exist and grow to the long-term size desired. Meaningful cash-flow projections require meaningful information that can be used as the basis for decisions.

At this point the Smith family has worked with contractors and has rough estimates of the costs of facilities to expand to 300 cows (Table 18-4). To complete the analysis, additional information relating to replacements, labor, and so on must be gathered.

TABLE 18-4 Initial facility cost estimates for a new 300-cow dairy.

300-stall freestall barn	$450,000
D-8 parlor in the existing barn	80,000
Feed storage	40,000
Manure storage	50,000
Miscellaneous equipment, excavation, and so on	80,000
Total facility cost	$700,000

Additional Animals and Herd Replacements

As herds expand there is often a need to add animals. This can be done by buying cows or heifers. If cows are purchased, extreme caution should be paid to their udder health, and they should be tested for common diseases. If heifers are purchased, they should have known ancestry and be trained to use freestalls. If you will be using self-locking manger stalls, selecting animals accustomed to them is also important. Whatever type of animal is purchased, remember to work with your veterinarian to establish a sound **vaccination program** for both new and existing animals. Keep these animals separated from the remainder of the herd for 30 days if possible.

Currently, the price of animals has increased dramatically. Monitor the heifer market to ensure values used for cash-flow projections accurately reflect current costs. Heifer breeding dates are very important. Producers have found that buying animals without knowing exact breeding dates results in animals calving later than expected, which increases feed cost and can lead to overconditioned heifers and excessive calving problems.

The Smith family arranged to have 50 springing heifers delivered in each of the first four months after startup, at an agreed-upon price of $1,500 per animal. They also used a computer model, based on herd size and expected cull rates, to determine that they would need an additional 100 head of heifers per year to cover replacement needs. Adding the cost of facility changes ($700,000) and the cost of additional animals ($300,000), the Smiths feel that they are within their $1,000,000 budget, and that further evaluation of the expected profitability of the proposed expansion is justified. To do this, they used figures from their past cost of production and verified them with benchmark database values to ensure that their numbers were representative of the industry. Table 18-5 shows average milk production cost per cow for 47 Wisconsin herds of more than 250 cows.

TABLE 18-5 Average milk production costs per cow in 2000 for 47 Wisconsin herds with more than 250 cows.

Purchased feed	$659
Milk hauling	46
Gas and oil	65
Insurance	26
Milk marketing	36
Repairs	138
Supplies	91
Taxes	27
Utilities	50
Veterinarian and medicine	106
Breeding	28
Other	831
Total basis cost	$2,103

Source: Frank, 2001.

Enterprises and Support Services

It is difficult to determine which enterprises will yield the greatest returns since both profit *and* risk removal are important considerations. You may be able to have your heifers raised less expensively by someone else, but that may increase the risk of introducing a disease into your herd. You may be able to have someone plant and harvest your crops, but will the work be performed on a timely basis? Because of the importance of these decisions, you must thoroughly evaluate your options and be comfortable that contracting support services is correct for your operation.

After reviewing their options, the Smiths decided to have their heifers custom raised and to hire a custom cropper to plant and harvest their crops.

Employees

If you do not currently have many employees, talk to other dairy producers in your area to determine the best source of employees and the normal salary expected.

Since the Smiths wanted a modern parlor, no heifers to raise, and custom-croppers planting their crops, they felt that they could cover the workload with the four full-time owners, their children, and two full-time employees. They already had one employee and knew a dependable neighbor who would fit into the operation.

Record Keeping and Data Collection

Modern dairies must maintain thorough financial and production record-keeping systems. The costs for computers, data collection devices, and support services are necessary for a meaningful project feasibility study. Both **initial costs** and **ongoing support costs** must be considered. Consulting with producers who have used automatic milk-recording and activity-monitoring devices, for example, may be helpful in determining the feasibility and costs of such technology.

The Smiths decided to upgrade their existing computer system and purchased a new financial record-keeping system, but decided to continue to use DHI to record animal production.

Contractors and Construction

Finding builders who have experience building modern dairies, and have the ability to deliver a high-quality product on time and under budget, is critical to the success of a major project. Talk with other producers who have recently expanded, and get their recommendations. Obtain firm bids on all aspects of the construction project (and on the related equipment) to ensure that the project does not exceed projected costs. Cost overruns of 10–15 percent are common with major construction projects, but much higher cost overruns have been experienced. To prevent cost overruns, get firm contractor bids, do not expand the scope of the project after contracts are let, and try to identify and budget for all the miscellaneous equipment and services that will be needed.

It is very important to decide early in the planning process who will be responsible for the project. Will you act as your own **general contractor,** or will you hire a general contractor or **construction manager** to manage the project? Be cautious if you plan to be your own general contractor. Do you have the experience, ability, and time needed? Can you afford to take the time and focus away from your current operation to perform this role?

Remember that it normally requires at least four hours per day during the construction phase to make the day-to-day decisions for subcontractors and to monitor the project. Since the Smiths did not have experience managing such a major construction project, they hired a general contractor (who had been recommended by a neighbor) to oversee the project.

Startup Needs

To make meaningful financial projections, you must know the amount of each resource needed by the dairy. Knowing the startup needs and having these resources available is critical to the success of the dairy. Often feed storage is constructed first and is filled as other dairy components are built. Additional animals may be purchased early and bred to calve when the dairy modernization project is completed. Additional labor needs should be identified and people hired and trained.

The Smiths used their existing records to determine the amount of inputs required per cow for their existing operation and estimated future needs based on these values.

Project Schedule

Plan construction to coincide with desirable local weather conditions. In the northern part of the United States, building should be completed and startup accomplished before severe winter weather arrives. If more than one contractor will be involved, consider having contractor meetings in which the participants can discuss schedules and completion dates. Written contracts should specify completion dates and penalties for not meeting them. Remember, when contracting, to have written lien waivers with each supplier of materials.

Since the Smiths live in a northern state where winter conditions complicate construction and adaptation of animals, they decided to build in the spring and to populate the facility in late summer. To avoid spring mud and to allow their builder to start early in the spring, they arranged to have excavation done the previous fall.

Contingency Planning

Planning is the process by which all of the resources needed to accomplish a goal are identified. Time and cost estimates are based on these assumptions, and total cost and completion times are derived from them. Since things do not always happen as expected, you must evaluate each portion of the plan, consider what could go wrong, and then formulate plans to cover these contingencies.

During the planning process, the Smiths felt that the three problems that could have the greatest potential effect on their expansion would be (1) not having the facilities completed on time, (2) insufficient feed because of a crop failure, or (3) a shortage of labor during their transition phase. To prepare for these possibilities, they made the following contingency plans. First, they planned (and contracted) to have the facilities completed during the summer so that if there was a delay, the extra cows could be housed outside and milked in the temporary milking facility that they would be using as their old barn was being converted. Second, they contacted local feed suppliers to reassure themselves that extra feed could be purchased if necessary. Third, they contacted several local high school students to determine if they would be able to help if needed. They hired two of them on a part-time basis to begin training them, in case they were needed later. They also discussed the impact of these provisions with their lender so that he was familiar with the situation and would support their needs.

Milk Production

The milk-production level of an expanding dairy can vary depending on many factors. To estimate the level of production to expect, consider cow comfort, magnitude of expansion, milking frequency, and genetic differences before and after the expansion. Adding a large number of heifers of unknown genetic quality will have a tendency to reduce production levels; increased milking frequency, better cow comfort, and a better feeding program, on the other hand, tend to increase production. Table 18-6 shows the average DHI herd milk production level of herds for expanded dairies that responded to the 1999 Wisconsin Dairy Modernization Project survey. On the average, producers were able to maintain levels of milk production and numbers of days open the year of expansion. The year after expansion, producers experienced significant increases in both values. To be conservative when doing financial analysis, base income projections on a 10 percent drop in average milk per cow the first year, unless circumstances justify a different assumption.

TABLE 18-6 Effect on production during expansion.

	Average DHI Milk	Average Days Open
Year before expansion	20,729	128
Year of expansion	20,803	129
Year after expansion	21,236	139

Since the Smith herd had maintained a 20,000-lb RHA for several years on 2X milking, with housing that was not conducive to high milk-production levels, they felt that they could maintain or actually increase production in the new facilities, but to err on the safe side they made all their financial projections based on 18,000 lbs per cow per year.

Putting It All Together

The steps just summarized, if done correctly, should yield the information needed to test the feasibility of your plan. Computer-planning programs can generate cash-flow, profitability, and efficiency estimates.

Table 18-7 shows the results of the feasibility study done to evaluate the Smiths' long-term plan to milk 600 cows. They assumed that they would receive an average of $13 per cwt and 70 lbs of milk per cow per day for the first seven years after expansion. The computer program allowed them to start production at a lower level and increase it over time. This technique allowed them to determine cash-flow requirements on a month-by-month basis as they grew their herd. The Smiths used return-on-investment as their measure of profitability and did sensitivity analysis to see the effect of changes to milk production or milk price levels. Notice how return would drop from 7.6 percent to 2.0 percent if there were a 10 percent drop in milk production. Considering these results, they were comfortable with their long-term plan and began to evaluate the feasibility of going to 300 cows during their initial expansion phase.

TABLE 18-7 Projected return on investment for a 600-cow dairy.

	10% Less	Expected Price ($13/cwt)	10% More
10% less (63#)	–3.6%	2.0%	7.1%
Expected milk (70#)	2.1%	7.6%	13.0%
10% more (77#)	7.0%	12.9%	18.9%

TABLE 18-8 Financial analysis results, first phase, expansion to 300 cows.

Financial Factor	Value
Cash Flow*	$79,776/yr
Profitability—Return on investment (ROI)*	8.5%
—Return on assets (ROA)*	4.6%
Solvency—Initial equity %	37%
—Initial debt per cow	$4,162
Efficiency—Initial investment per cow	$5,278
—Cows per FTE	61
—Pounds milk per FTE*	1,351,449

*Seven-year average values

Table 18-8 shows the results of the initial financial analysis of the first phase of the Smiths' expansion project. Although the financial analysis was several pages in length, they selected some key values to evaluate the feasibility of the change. Expected cash flow is important, to determine if sufficient income will be generated to cover the bills (this is especially true the first two or three years after a major expansion). Month-by-month values should also be generated to determine and plan for income shortfalls. Since profitability is the key to the **long-term viability** of a business, it is important to select measures such as return on assets and return on investment in evaluating the proposed change. The solvency figures shown in the table are factors that lenders consider when making a loan and reflect the financial strength of the operation. Efficiency figures should quantify the efficiency of the operation and can be used to compare operations.

In this case, the Smiths needed to reevaluate their labor assumptions to be sure that six full-time employees (FTEs) would be sufficient. The key financial factors with which to evaluate a proposed change can vary depending on your objective.

SUMMARY

The U.S. dairy industry has been undergoing a major change, characterized by fewer but larger dairies. This change has been driven by new technologies that allow managers to successfully manage larger dairy herds. Producers are moving from a traditional dairy system, in which cows are housed and milked in a stall barn, to a system that may include intensive rotational grazing, freestall housing, TMR, milking parlors, robotic milkers, and so on. No single system will be best for everyone; therefore, it is important for producers to understand available options and evaluate the merits of each for their operations.

Modernization of existing dairies often requires major changes. Planning is critical to success and should not be rushed. It will take a typical farm family one to three years to formalize plans. Setting goals, getting buy-in from family members, visiting existing dairies, determining management objectives, selecting facility designs, getting realistic cost estimates, testing strategies, doing feasibility evaluations, locating resources, hiring contractors, and so on, take a lot of time. Knowing the pitfalls experienced by other producers—and developing plans to avoid them—is critical to success. Knowing where the industry is going and making management decisions that allow your operation to compete in the long term is equally important. Not everyone needs to have a large herd; the important questions are whether your operation is sustainable in the long term, and whether it provides the kind of returns and working conditions that you and your employees desire. If changes are needed, remember that direction is often more important than speed. Making several small changes often requires less initial capital and may be less risky than building an all-new facility. Modernization is not about keeping up with the Joneses; it is about positioning your operation to be competitive in the long term.

CHAPTER REVIEW

1. Why is no single modernization plan best for every producer?
2. List the five decisions that most affect a dairy's modernization plan, once financial limitations and opportunities are known.
3. List the steps in conducting a dairy-modernization feasibility study.
4. How long does it typically take to formalize modernization plans?
5. Define "contingency plan."

REFERENCES

Frank, G. (2001). *Milk production costs in 2000 on selected Wisconsin dairy farms.* Madison: University of Wisconsin.

Palmer, R. W., & Bewley, J. (2000). *The 1999 Wisconsin dairy modernization project—Final results report.* Madison: University of Wisconsin, Dairy Science Department.

Chapter 19

Expansion Examples

KEY TERMS

elevated platform
gray water

Objectives

After completing the study of this chapter, you should be able to

- visualize affordable retrofitting methods.
- develop creative yet viable ideas for the reuse of existing structures.
- incorporate these ideas into dairy modernization plans.

Throughout this book, pictures of new dairies were often used as examples to illustrate design features of modern dairies. Many producers do not wish to build all new facilities, or would prefer lower-cost options. The figures in this chapter show examples of existing dairies that modernized operations in recent years, using lower-cost options. Most producers in these dairies had 45–60 cows before expansion and 100–150 cows after expansion.

Example 1: Low-Cost, Back-Out Flat Parlor in Old Barn and New Freestall Barn

Figure 19-1 shows a D-10 flat-barn parlor built in the end of an old dairy barn near an existing milkhouse. Gutters were filled with concrete. Milking units are removed from the barn after each milking; three existing stalls are used per milk machine unit. Cows are loaded in the first and third stalls, and the unit is mounted in a protected cage in the second stall between the two cow stalls. This is repeated five times on each side of the barn, with a total of five units mounted per side; fifteen existing stalls on each side of the barn are used. Cows are loaded in batches. The milking machine units are placed on even-sided (or odd-sided) stalls first. When these cows are finished milking, the machine operator moves the unit to the cow on the other side. After all animals of a type (odd or even) have had their milking unit removed, they are released in a group, they back out of their stalls, leave the barn, and a new batch of animals is loaded. This process continues as the operator alternates between even and odd sides. Sawdust is sprinkled on the floor to capture any manure or urine moisture, and cow-access lanes are

FIGURE 19-1 Low-cost, back-out flat parlor in the end of an old dairy barn.

FIGURE 19-2 Gang locking mechanism of existing stanchions.

FIGURE 19-3 Return lane.

manually scraped periodically. Figure 19-2 shows the gang-locking mechanism of the existing stanchions, which was modified to allow cows 1, 3, 5, 7, and 9 to be locked and released with one lever, and cows 2, 4, 6, 8, and 10 to be locked and released with the second lever. Figure 19-3 shows the remainder of the existing dairy barn, which had stalls removed, gutters filled, and fences made. Two-thirds of the barn is used to hold animals waiting to be milked, and the other third as a return lane to the new freestall barn.

Example 2: Walk-Through Flat Parlor in Old Barn and New Freestall Barn

Figure 19-4 shows a D-4 walk-through flat-barn parlor that was built in an old dairy barn. Cows step onto an elevated platform, are milked, then exit through the front head gate. This type of parlor has one milking machine unit per stall, so when a cow has finished milking it is necessary to complete the milking process, have the cow exit, get the next cow into the stall, and attach the milking unit. Since cows need not back out of the stalls, the animal platform can be 12 inches or higher (**elevated platforms** allow better machine-operator routines).

Figure 19-5 shows a cow exiting the front of its stall after being milked in a walk-through flat-barn parlor. As soon as a cow leaves the stall, the front gate closes and another cow may enter the stall to be milked. This parlor type may be constructed with one or two milk stalls per milking unit. If the parlor has one stall per unit, then it is very important to remove the cow, load a second cow, and quickly attach the unit if maximum parlor throughput is to be achieved. If the parlor has

FIGURE 19-4 D-4 walk-through flat-barn parlor, built in an old dairy barn.

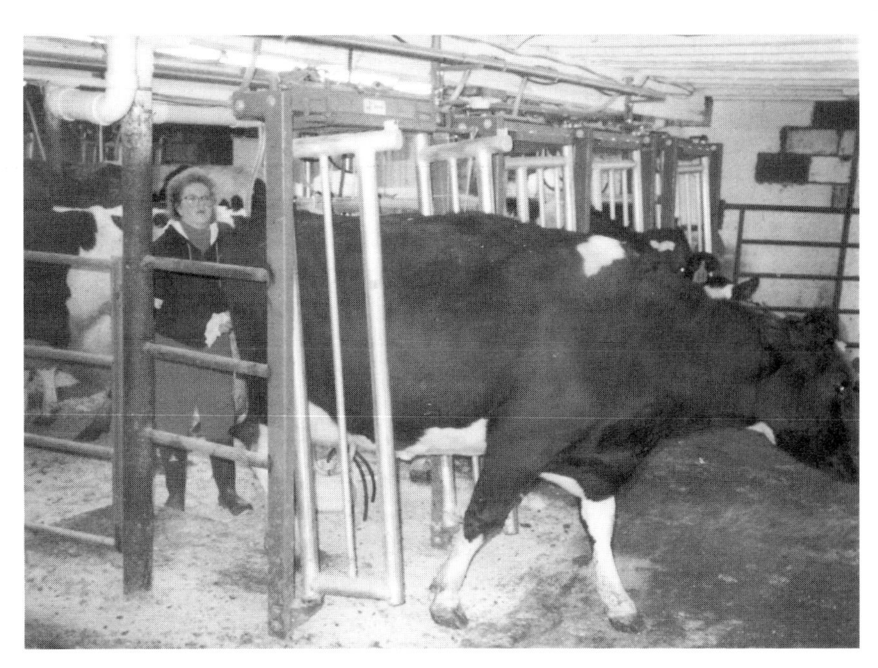

FIGURE 19-5 Cow exiting a walk-through flat-barn parlor.

FIGURE 19-6 A milk stool mounted in a flat-barn milking parlor.

two stalls per unit, the time to empty and load the stall is of less importance since the milking unit is busy milking a cow in the other stall. Often a milk stool is mounted in a flat-barn parlor to allow the machine operator to sit as cows are prepped and milking units are attached (Figure 19-6).

FIGURE 19-7 D-4 pit parlor placed in the end of an old dairy barn.

Example 3: D-4 Pit Parlor in Old Barn and New Freestall Barn

This producer built a D-4 pit parlor (Figure 19-7) in the end of an old dairy barn near an existing milkhouse. Some stanchions were removed to provide space for the new parlor, but the remainder of the stanchions were retained and used for other purposes.

Manure gutters were covered with grates, and the cow access lane between rows of stalls is used as a holding area for cows waiting to be milked (Figure 19-8). Gates were mounted, which form the sides of the holding area. These gates are moved and the old stanchions used to handle animals between milkings. A new 120-cow freestall barn was constructed near this existing barn, and cows walk outside to be milked and to return to their housing.

Figure 19-9 shows the feed alley in front of the old stanchions inside an old dairy barn that was converted to house a low-cost pit parlor. The original feed manager was split to provide a parlor return lane for animals after milking, and space for feeding animals locked in the stanchions.

Example 4: D-26 Swing Parlor in Old Barn

This producer built a D-26 swing parlor in the end of an old dairy barn (Figure 19-10). It is normally operated by two people and reportedly can milk close to 200 cows per hour. Manure splattering can be a concern with

FIGURE 19-8 Holding area.

FIGURE 19-9 Return lane from a low-cost D-4 pit parlor and a feed alley used to feed cows restrained in the stanchions shown.

this type of parlor because there are no manure shields to protect the operator. This low-cost parlor had a holding area located at one end of the barn, used to hold cows waiting to be milked (Figure 19-11). The barn was divided to provide space for the parlor, a palpation rail, and a return lane.

FIGURE 19-10 D-26 pit parlor with swing parlor equipment.

FIGURE 19-11 Holding area for low-cost D-26 swing parlor.

After being milked, cows were retained in the palpation rail if they were to be bred or treated, or exited the building via the return lane. This parlor was not heated, so condensation formed when the warm water used during milking was exposed to the cold concrete.

Figure 19-12 shows a cow being milked in this parlor. An inexpensive stall design was selected to minimize capital cost, but this allowed manure to splatter because no manure shields were provided. Milking without manure

FIGURE 19-12 Cow being milked in low-cost D-26 swing parlor.

FIGURE 19-13 Washing manure from a low-cost swing parlor with a high-pressure water hose.

shields is acceptable to many producers for one- to two-hour shifts but may not be acceptable for longer shifts.

Low-cost parlors are often equipped with high-pressure water hoses to periodically clean the parlor platform (Figure 19-13). Concrete should be sloped to channel the water and manure mixture (**gray water**) to proper

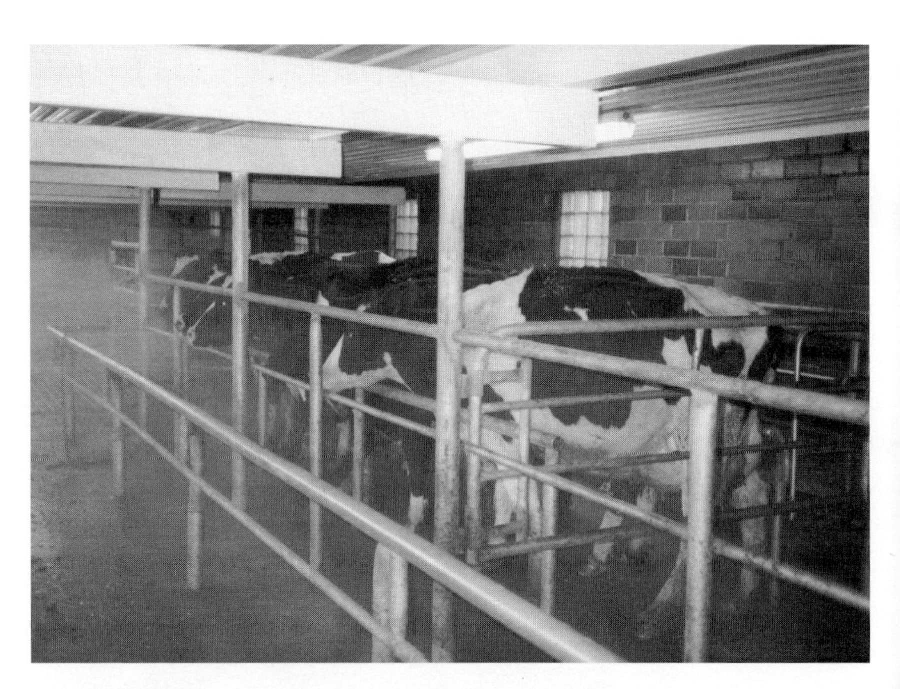

FIGURE 19-14 Cows retained in a palpation rail, waiting to be bred or treated.

storage. As cows exit the milking parlor, they may be sorted into a palpation rail where they can be bred or treated before returning to their home pen (Figure 19-14). Cows not selected can pass and exit the back of the barn to return to their housing.

Example 5: New, Low-Cost Facility with D-6 Parlor and Four-Row Freestall Barn

The new dairy facility in Figure 19-15 was designed to provide comfortable working conditions at a minimal cost. A T-style design was selected. Both the milk storage and equipment rooms provide for the essentials only. A bathroom and space for a desk in the equipment room support the needs of the operator.

Figure 19-16 shows the holding area and return lane from a new low-cost dairy facility. Cows exit the D-6 parlor on a single return lane, which requires less building width. The holding area connects to the freestall barn and is equipped with an inexpensive crowd gate. A small treatment area was provided off the return lane to catch and treat animals.

Figure 19-17 shows the inside of the D-6 herringbone parlor that was built as part of this low cost dairy complex. This single-return-lane design requires cows on one side of the parlor to walk around the front of the

FIGURE 19-15 New, low-cost parlor and freestall complex, arranged in a T-style configuration.

FIGURE 19-16 Holding area with single return lane and low-cost crowd gate.

parlor to exit via the single return lane. Milk machine operators must walk across the cow exit lane and then down into the parlor pit. Cows often defecate as they leave the parlor, making it difficult for milk machine operators to enter the pit without encountering manure.

Figure 19-18 shows the four-row barn that was constructed as part of this new, low-cost dairy facility. It is constructed with wood rafters, which are normally less expensive but provide a place for birds to perch, which can be a problem in many areas of the country. It is equipped with

FIGURE 19-17 Inside a D-6 herringbone parlor.

FIGURE 19-18 A four-row drive-through freestall barn with wood rafters; only a portion of barn is equipped with self-locks.

a feed rail for most of the feed space, and only a few self-locks at the end. The area with the self-locks can be made into a pen by moving two gates. Cows can be sorted into this pen, and animals restrained for treatment with the self-locks. Freestalls utilize sand bedding to minimize capital cost.

Example 6: New, Low-Cost Facility with D-6 Parlor and Two-Row Freestall Barn

Figure 19-19 shows a new, low-cost freestall dairy facility. It is a T-style facility, with a D-6 herringbone parlor and a two-row freestall barn. The barn is narrow, which makes it inexpensive to build and easy to ventilate. It contains two pens of cows, and can be expanded later to four pens by adding onto the end shown. The herd can be expanded again by adding a second barn parallel to the first. This operation is owned by a family that had milked 45 cows in a tie-stall barn, but now milks 110 cows with the same family labor.

Figure 19-20 shows the interior of the two-row freestall barn shown in Figure 19-19. This freestall barn has one row of stalls on each side and a

FIGURE 19-19 New, low-cost freestall facility with two-row barn and D-6 pit parlor.

FIGURE 19-20 Interior of two-row drive-through freestall barn.

FIGURE 19-21 Interior of low-cost T-style freestall facility.

short row of stalls at the end of each pen. The number of freestalls in the short row is determined by the desired amount of feed space per cow. In this case, there are three stalls on the long rows for each stall on the short rows, providing two feet of feed manger space for each cow. Notice how the stalls on the far end extend into the area used for feed in the remainder of the barn. The space between the freestalls on the far end of the barn is being used to store a feed wagon between feedings. Waterers are placed at the ends of the feed manger. This barn design has only one manure alley per side, so cows must be removed when manure is removed, or alley scrapers must be used.

Figure 19-21 shows the holding area for cows waiting to enter the parlor, a single return lane, and individual animal pens near this return lane. Cows needing special attention can be diverted from the return lane into the pens for treatment. The window shown (Figure 19-21) allows the manager to watch cows in the pens (Figure 19-22) from his office located behind the window. Access to the office and equipment room is via a door to the right of the window shown. A walkway extends from the door into the equipment and office area past the pens and can be used to deliver feed to animals in the pens. Pens are divided by gates that can be moved for manure removal. Designs with individual pens located on the side of the parlor holding area are not recommended, because the location makes it difficult to automate feed delivery and manure removal. Animals located near where cows enter or exit the parlor can also impede the flow of animals into or away from the parlor.

FIGURE 19-22 Space in front of individual pens to provide feed and water to animals.

FIGURE 19-23 Inexpensive D-12 swing parlor facility.

Example 7: New D-12 Swing Parlor Built by Grazier

Figure 19-23 shows a new dairy facility that contains a D-12 swing parlor used by a grazier. The facility is located near paddocks and was inexpensive to build. The holding area is enclosed and contains an inexpensive crowd gate to help move cows into the parlor.

Figure 19-24 shows the interior view of a new D-12 swing parlor that incorporated low-cost stalls and inexpensive building materials to keep the cost to a minimum. Although this design was chosen by a grazier, it would be equally effective in supporting the needs of producers with confinement operations who wish to minimize milking-facility costs.

FIGURE 19-24 Interior view of new D-12 swing parlor.

FIGURE 19-25 New dairy facility with a freestall barn and robotic milking equipment.

Example 8: Two-Box Robot Milker and New Four-Row Freestall Barn

Robotic milking systems are becoming more common in the United States. Figure 19-25 shows a new dairy facility with a freestall barn, two robotic milker boxes, an animal treatment area, and a milk room.

Each robot box (robotic milk stall) is equipped with a one-way gate to direct cows into the stall, where they are fed grain and milked (Figure 19-26).

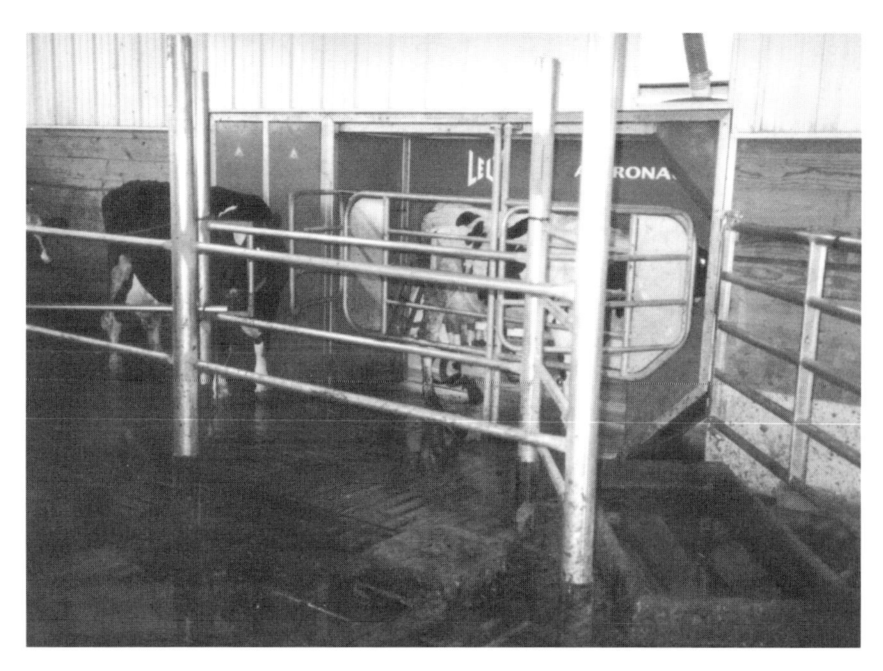

FIGURE 19-26 Robotic milk stall with one cow being milked and one waiting to be milked.

FIGURE 19-27 One-way gate that allows cows to leave the feed manger area and return to the freestall portion of the barn.

After animals leave the robot box, they exit onto the feed alley, where they are normally fed TMR rations.

After animals have finished eating the ration provided at the feed manger, they can enter the freestall portion of the barn via a one-way gate (Figure 19-27). After cows have been conditioned and learn to access the milk stall, the one-way gate system can be removed. Experience has shown that removal of the one-way gate results in increased feed consumption and milk production.

CHAPTER REVIEW

1. In your opinion, what facilities are easiest to retrofit in an existing dairy operation? Explain your answer and give examples.
2. In your opinion, what facilities are most difficult to retrofit in an existing dairy operation? Explain your answer and give examples.
3. If a dairy is modernizing but has limited means, what new facilities would you plan to build or retrofit in the first phase of expansion? Why?

Glossary

2X (2X milking) An abbreviation denoting that cows are milked twice per day.

3-basin storage system A manure storage system that also operates as a separator, with the first basin separating out heavier solids, the second settling out sediment, and the third holding the separated liquid used to fill the flush tanks.

3X (3X milking) An abbreviation denoting that cows are milked three times per day.

4X (4X Milking) An abbreviation denoting that cows are milked four times per day.

accounts payable The list of bills to be paid (to vendors, etc.).

added water Water added to milk to increase its volume—an illegal practice.

affiliative A leadership style in which people come first and tasks second.

AI progeny-tested bull A bull whose genetic merit has been proven using the production records of daughters produced using artifical insemination (AI).

AI sire A sire whose semen is available through AI.

AI young sire A young bull whose semen is available through AI.

alley scraper An automatic device used to scrape manure out of alleys and into reception pits.

alley scraper drive unit The functional piece that motorizes an automatic alley scraper.

AMS (automated milking system) In this system, a cow enters the milking stall without manual intervention, a computerized milker cleans the cow's udder, attaches the milking unit, removes the unit when the cow is finished being milked, disinfects the cow's teats, releases the cow, and prepares to milk another animal. Also known as *robotic milking system.*

animal flow The physical pathways animals take to move through a dairy site.

animal handling activities The activities required to sort and restrain animals for treatment.

animal movement information Information regarding the movement of animals from one location to another on a dairy.

animal replacement systems The methods used to acquire replacement animals for a herd.

antimicrobial drug residues Residues of drugs used to kill bacteria and other organisms.

assets The entire property of a person or business; elements that are required to run a dairy.

attach To place a milking machine onto a cow's teats.

authoritative A leadership style in which subordinates are not invited to participate in the formulation of management ideas.

automated parlor A parlor in which automated equipment is used to perform part of the milking process (milking machine claw removal, milk recording, etc.).

automatic alley scraper An automatic device used to scrape manure out of alleys and into reception pits.

automatic detachers Equipment designed to remove the milking machine claw from a cow's teats when a sensor indicates that the milking process should be terminated because of low milk flow.

automatic sort gate An electronic gate that automatically sorts animals by reading their electronic identification tags.

autopsy An examination of a carcass to determine the cause of death.

auto-tandem parlor An individualized parlor in which stalls are placed end-to-end. Cows are loaded individually rather than in groups; when one cow leaves a milk stall, another cow is free to enter.

average annual total cost The cumulative cost of an activity or product over the course of one year.

back pen An animal pen that does not have direct access to a common lane or alley.

back-out flat-barn parlor A parlor installed in an old barn, in which cows must back out of stalls after milking to exit.

balance sheet A document that lists all assets, liabilities, and the current net worth of a person or company.

bare cropland Land that can be used to raise crops.

barn cleaner A mechanical piece of equipment used for cleaning a barn in a traditional dairy management system.

barn orientation The physical orientation of a barn on a dairy site.

basement (subway) parlor A parlor in which the milk meters and some of the milking equipment are placed below the cows or the operators.

bedded pack A housing pen in which animals are free to lie wherever they wish, and whose floor is covered with some type of bedding material.

benchmark database A database with statistics that reflect regional averages for production levels, pay, and so on.

biological subsystem All elements of a dairy that are related to animal health.

biosecurity The monitoring of all existing and incoming animals for disease, and the use of disease prevention techniques.

body capacity An animal's depth, length, and width, which serve as an indicator of the animal's ability to consume feed.

bonus A financial incentive given to an employee as compensation for excellent work.

bottom-line profitability An informal term for the profitability of an enterprise or operation, based on its income generation and operating costs.

brisket board A board installed in the front of a freestall that is used to position the cow correctly when lying in the stall.

bST injections Injections of bovine somatotropin, a growth hormone designed to increase milk yields.

bucket milker A traditional milking unit in which the milk removed from the cow is stored in the milk receiver (bucket) portion of the unit.

bulk tank The common tank used to house milk produced by a herd.

bunker A horizontal trench, lined with concrete or some other sidewall material, in which silage is stored.

business goals The objectives of the business as determined by the owners.

business plan A summary of interested parties' ownership, management, and operational goals once modernization is complete.

business structure The legal structure of an operation, often affecting taxation and the transfer of ownership.

business planning process The process of planning business operations.

butt pan An item designed to catch manure in the back of a milking parlor stall, to prevent it from falling into the stall or onto milking equipment.

buy-sell agreements Agreements that put limits on the transfer of ownership of a business.

buy-sell contracts Contracts in which goods are automatically purchased when the market price is within a certain range.

BVD Bovine virus diarrhea, a viral disease.

cash flow A summary of the money spent and received by a dairy.

cash wages Payment given to employees as compensation for work.

cash-flow projections Predictions of the income and expenses expected for a specific period in the future of a business.

catch lane A pathway created out of gates or fences, used to facilitate the restraint of animals.

catch pen A temporary holding pen for animals that have been sorted.

cattle guard A device made of pipes, placed at ground level near entrances to alleys or pens and designed to keep cattle from entering.

cattle shades Shades designed to provide protection from direct sun for cows, feed, and waterers.

C-corporation The most common corporate entity, formed and endowed by law with the rights and liabilities of an individual. Profits are taxed as business profits, and dividends taxed as personal income, which can result in some profits being taxed twice. See also *S-corporation.*

center alley The common lane through the center of the barn that is used to deliver feed to animals on both sides of the barn.

chart-of-accounts A list that categorizes financial data, used for summarization of income, expense, and so forth for a given enterprise or business.

chronic staph A form of the mastitis-causing staphylococcus organism that is difficult to eliminate and tends to recur.

clay-lined storage basin An in-ground storage basin for manure, lined with clay.

clinical mastitis The degree of mastitis infection (inflammation of the udder) that can be physically observed (as soreness or hardening of the udder, or thick flakes in the milk produced by the cow).

cluster spacing The amount of space a milking parlor stall consumes along the pit wall, as measured between the centers of adjacent milker claws.

coaching A leadership style in which a manager develops workers as subordinates; the act of developing employees by allowing them to try their hand at a particular task.

coercive A leadership style in which a manager expects immediate compliance from workers.

coliform count (CC) A measurement of fecal bacteria in milk.

color rendition The ability of a lighting system to allow people to accurately discriminate color values.

commodity storage An area in which commodities (by-products), grain, or concentrates are stored.

company policies Principles that dictate how a company will be run.

company rules Guidelines for employee conduct.

compensation Payment for work completed.

composting Blending manure with a source of carbon and allowing it to decay and to create soil.

connector barn A covered alleyway that connects a freestall barn to another freestall barn or a milking parlor.

construction manager A person given the responsibility to manage the construction or modification of a dairy facility.

contact time The actual time spent manipulating and touching a cow's teats to stimulate oxytocin release.

contagious bug A disease transferred from one cow to another as a result of physical contact.

contagious organisms Organisms that can be transferred from one animal to another by contact, or via contaminated food and water.

contingency plan A backup plan that details steps to be taken to remedy negative effects of unforeseen events or circumstances.

contracted forage Forage for which a supplier and owner have agreed upon terms of sale.

contracting Developing a binding agreement between two or more parties to provide a product or service to the dairy.

conventional dairy barn A traditional dairy barn using older technology for housing and milking.

cooperative A corporate structure in which each member has an equally weighted vote in management decisions.

corral An outdoor pen used in warm climates to house many animals.

corrective action Action taken to punish an employee for rules that are broken.

cover forage The process of covering forage so that it does not spoil.

cow health The current health status and treatment of a cow.

cow lanes The physical pathways used by cattle to move through a dairy site.

cow milk composition A definition of the components of milk (fat, protein, lactose, etc.).

cow milk production A cow's current and historical milk production levels.

cow pedometer readings Automated measurement of a cow's activity using a device attached to the animal's leg.

cow pusher A person who brings pens of cows to the milking parlor and returns them to their home pen.

cow reproduction All activities relating to the process of getting cows to reproduce.

cows per hour (cph) The number of cows milked in one hour.

cows per labor hour (cplh) The average number of cows milked per hour of worker labor used.

cross channel A concrete channel used to collect manure removed from freestall barn alleys.

crowd gate Movable gates located in the holding area near a milking parlor, used to encourage cows to enter the parlor.

culling rate The average annual percentage of animals eliminated from the herd.

custom breeder An AI technician who contracts with a producer to perform all the breeding-related activities on a dairy for a specified rate.

custom cropper A supplier who contracts with a producer to do specific cropping-related activities for an established rate.

custom heifer raiser A supplier who agrees to raises heifers for a producer at an established rate.

custom manure hauler A contractor who specializes in hauling and spreading manure for dariy producers.

custom milker A non-employee who contracts to milk cows for a fixed rate.

custom operator A supplier who agrees to perform some crop-related activity (planting, harvesting, etc.) for a fixed rate.

D-# parlor A designation of milking parlor size, in which "D" represents "double" and "#" represents the number of stalls per side. A D-8 parlor would have a total of 16 milk stalls.

dairy herd management systems (DHM) Computerized information systems used by dairies to record an animal's identification and reproductive, health, and milk-production histories.

dairy records processing centers (DRPCs) Computer-processing centers that collect and process data gathered at individual dairies for on-farm management use and to support industry-related evaluations.

days-in-milk The number of days in a lactation during which a cow produces milk.

debt-per-cow A measure of solvency calculated by dividing the total liabilities of an operation by the number of cows in that operation.

delegating Assigning tasks to employees, with evaluation following.

democratic A leadership style in which decisions are made in a participative manner.

descriptive information Summary information regarding the performance of an operation.

DHI (dairy herd improvement) A dairy industry data-collection and record-processing service that has been the standard milk-recording system in the United States for many years.

DHIA Dairy Herd Improvement Association.

diagnostic information Information generated by comparing descriptive information and external standards in order to diagnose problems in an operation.

diffuser The unit through which water is released to flush manure alleys in a freestall barn with a water-flush system.

direct cost Cost incurred as a direct result of an event.

directing Informing and showing employees how to do a new task with which they are not familiar.

dismissal Firing an employee or being fired.

double-# parlor Designation of milking parlor size, in which "double" implies two sides and "#" represents the number of stalls per side. A double-8 parlor would have a total of 16 milk stalls.

drive unit The functional piece that motorizes an automatic alley scraper.

drive-by barn A barn designed for drive-by feeding.

drive-by feeding A method of feeding cattle in freestall barns with feed mangers on the outside of the barn.

drive-through barn A barn designed for drive-through feeding.

drive-through feeding A method of feeding cattle in freestall barns with feed mangers inside the barn.

drover lane A cow lane used to take groups of animals from one location to another.

dry matter (DM) basis The normal method of evaluating a feed's nutrient content, based on determining the nutrient density of the feed after all moisture has been removed.

drylot system A dairy housing system often utilized in warmer climates, in which cattle live outdoors in large pens rather than in barns.

dystocia Calving problems, usually with heifers, normally caused by a disproportion between the size of the calf and the dam (i.e., the calf is too large to pass through the birth canal).

early fresh cow A cow that has had a calf in the last few days.

economic subsystem All elements of a dairy related to its financial health and stability.

electronic dairy data collection (EDDC) systems Information management systems housed within individual dairies that electronically monitor a cow for milk production levels, activity, and so on.

elevated platform A platform raised above the floor of a facility, on which cows stand.

e-mail An electronic form of written communication.

employee handbook A document that explains all the essential elements, policies, and rules of a company.

employment-at-will A legal term that implies an employee can leave a job at any time and that employers can terminate employment at any time.

enterprise A portion of a business that can be thought of and managed as a separate business.

environment information Information regarding surrounding conditions (temperature, humidity, air quality, etc.).

equipment flow The physical pathways equipment must travel in order to maintain a dairy.

equipment monitoring Automatically checking equipment used on dairy (bulk-tank temperature, pulsation function, vacuum stability, etc.).

estate plan A plan that details how assets or interest in a business will be transferred to other parties upon the death of the owner(s).

estrus The time at which a cow is sexually excitable, will accept the male, and is capable of conceiving.

eave opening Openings along the sidewall of a barn, just below the roof, that allow for natural ventilation of the building.

farm foreclosure A bank or lender seizing a farm because of nonpayment of loans.

farming out An informal term for using contract services.

farmstead A dairy site and its components.

far-off dry cow A cow that has been dried off (milking terminated) several weeks before she is expected to have a calf (normally 21–60 days before expected calving).

feed-acquisition systems The methods used to raise or purchase feed for a herd.

feed inventory Information regarding the amount of feed available to the dairy.

feedline sprinkler A sprinkler system installed along a feed manger to cool cattle on hot days.

feed-mixing program feedback Information (from computer programs such as E-Z Feed) regarding loading and unloading of rations.

feed shrinkage Loss of feed weight due to moisture evaporation or fermentation after harvest.

feeding program Feeds used by a herd, and the different rations fed to different groups of animals.

feeding systems The methods used to feed a herd.

feeding waste Feed given but not consumed due to enviromental conditions (rain, wind, sun) or animal refusal.

feedline A feeding slab or manger that lines the edge of a corral.

feed-scraping equipment Equipment used to remove unconsumed feed from a manger.

fenceline feeding Delvering feed at a feedline.

fermentation shrink The volume lost as feed ferments into silage.

fermentation waste See *fermentation shrink*.

financial details All transactions dealing with purchases, sales, payments made, and payments received by an operation.

financial feasibility The determination that a business decision has the potential for financial success.

financial liability Responsibility for repayment of debts.

financial status Information regarding the financial health of an operation, including loan balances, accounts receivable, accounts payable, and projected income and expense streams.

flat feed storage Any feed storage system in which feed is stored in piles over a large horizontal plane rather than in vertical upright storage units.

flat-barn parlor A milking parlor that does not have a milker pit.

flexible-drag-hose system Equipment used to spread manure on fields. Manure is pumped from storage through a hose, to a unit that incorporates the manure into the soil.

flexible-membrane-lined system An in-ground storage basin for manure, lined with a flexible membrane fabric.

flush manure removal Using water to flush manure down an alley and into a reception pit.

foot bath A flat container filled with water and disinfectants that cows walk through to cleanse hooves and prevent infection.

forage dry matter (DM) The percentage of forage that remains if all the water is removed.

freestall barn A barn with pens that allow animals to move as they wish, and with freestalls for resting.

freestall divider A rail system that separates individual freestalls.

freshen When a cow has a calf.

front pen An animal pen with direct access to a common lane or alley that is in front of another pen of animals.

FTE (full-time equivalent) A measurement of labor usage based on the normal number of hours that an employee will work.

full prep A procedure used to prepare teats for milking that involves all of the udder-preparation steps recommended for proper hygiene.

general contractor A company or individual hired to be responsible for retrofitting or constructing the facilities in a dairy.

general partnership A business relationship in which at least two owners jointly own and are equally responsible for the business and all of its debts.

generic product A commodity (nondifferentiated) product.

geological homework The study of the soil, water, and environmental status of a potential or existing dairy site.

goal setting The process of defining the objectives of a business, activity, or individual.

goals Expectations set for a business, activity, or individual.

gravity channel A channel that uses elevation and gravity to carry manure from a reception pit to a storage facility.

gravity-flow manure handling system A system that takes advantage of elevation changes to move manure from a freestall to a storage tank.

gray water Water from a milking parlor that contains manure, soap, or disinfectant residues.

grazing system A dairy management system that relies upon pasture feeding.

grouping milking routine A milking parlor routine in which one operator performs all individual tasks of the milking procedure on a group of cows, then moves to the next group of cows once all tasks are complete.

hairy heel warts Digital dermatitis, a common lameness problem of dairy animals.

harvest waste The feed that is lost or discarded during the harvesting process.

HE (hundredweight equivalent of milk sold) An accounting method used to convert the income from nonmilk sources to the quantity of milk required to achieve the same income, used to standardize and compare different operations.

head milker A milker who is in charge during a milking shift.

head-lock A type of stanchion used to restrain an animal.

head-to-head barn A freestall barn designed with rows of cattle facing one another.

head-to-tail barn A freestall barn designed with one row of cattle facing the rear of the next row.

heat-stressed cattle Cattle that are suffering from an inability to dissipate sufficient heat to maintain normal body functions. Caused by excessive temperature, radiant energy, or relative humidity.

heifer A cow before she has her first calf.

herd components The average butterfat and protein percentages of the milk produced by a herd of cows.

herd culling history The historical level of and reasons for animals being removed from a herd.

herd inventories Current number of animals by type (milking, dry, and heifers) and by location.

herd milk level The average milk production level of cows in the herd.

herd production The actual quantity of milk shipped per day.

herd startup size The number of milking cows in a dairy at the time a new phase of expansion is initiated.

herd turnover The culling rate of a herd.

herdsperson A person who is responsible for the health and reproductive aspects of a herd of cows.

herringbone parlor A style of milking parlor in which stalls are constructed at a 45° angle to the parlor wall.

high-line A milk transport system in a milking parlor that carries milk from the cows to the bulk tank, located higher than the cows being milked.

holding area A pen in which cows are kept as a group before being moved into the parlor to be milked.

home-based animal handling systems A method of treating animals in the pen where they are normally housed, often using self-locks to restrain them.

horizontal silo A horizontal trench, lined on the sides with concrete or other materials, in which silage is stored.

housing capacity The total number of animals that can be housed at one time in a particular housing facility.

housing systems The type of facilities where animals are kept.

H-style A barn configuration in which two freestall barns have a parlor–holding area complex parallel to them and located between them.

HVLS (high-volume low-speed) fan Extremely large fans that are designed to move slowly, moving large quantities of air at a slow speed.

hybrid organization An operation with different business structures for different elements of the business.

implementation plan A plan that details how a particular goal will be achieved.

inbreeding The mating of closely related animals.

incentive payments Payments given to employees for achieving a particular goal.

income statement A statement that shows all incomes and expenses for a business for a specific period.

indexing Mechanical equipment in the front of a milk stall, used to move animals back in the stalls so they are positioned correctly to be milked.

indirect cost Costs incurred as an indirect result of an event.

individual calf hutch Individual housing units for pre-weaning calves.

initial cost The cost to purchase an asset.

Internet Electronic mail and World Wide Web services, accessed via a computer.

involuntary culling That portion of the animals leaving the herd that is not determined at the manager's discretion.

island waterer Water tanks located between and shared by two pens of animals.

job application An information form that potential employees fill out to apply for a job.

job description A written description of the essential elements associated with a specific role in a business.

job requirements A specific set of background experiences required of employees for a specific role in a business.

job summary A brief description of a job's purpose and tasks.

Johne's disease Paratuberculosis, a slow-growing, incurable infectious disease that affects the productivity and profitability of dairy cattle.

jumper bulls Bulls used to naturally service cows.

kill time The amount of time it takes a teat dip to kill bacteria present on the teats.

labor systems The systems required to properly manage labor and carry out all necessary tasks in a dairy.

labor use and compensation level Details on hours worked and wage rates paid.

laboratory pasteurized count (LPC) A measure of bacteria that survive pasteurization.

lactation cows Cows that are producing milk.

lagoon An outdoor pond or basin that holds manure water used for water-flush manure handling systems.

large modern confinement system A dairy management system that normally has several pens of cows, uses a milking parlor, and houses animals in freestall barns or drylots.

late-payment penalty A charge for paying a bill after it is due.

lease rate The proportion of the value of an asset paid for its use to another person or business that owns the asset.

lifts Mechanical devices used by workers to raise an animal if the animal cannot rise by itself.

limited liability corporation (LLC) A corporation in which all partners are responsible only for the amount invested or personally guaranteed.

limited partnership A business arrangement that has at least one general partner, who has business management responsibilities and full responsibility for debts, and at least one limited partner, who does not participate in management and has limited liability for debts.

liquidity The measure of a business's ability to meet its ongoing operational financial obligations.

loaded mile A mile driven by a custom hauler when his or her truck is loaded with goods being delivered.

long-term contracts Contracts that last for an extended period (usually for several years).

long-term goals Goals that are expected to be achieved over a long period.

long-term viability The determination that a business will be financially stable for a long period.

machine on-time The amount of time a milking machine is attached to a cow's udder.

mail-in service A service in which an operation's data and records are processed and analyzed at a remote location.

management grouping The process of grouping animals that have like needs.

management rail system A simple structure (normally constructed of pipes) that is used to restrain animals in a common area; also referred to as *palpation rail system.*

management system The set of philosophies and methods used to manage a dairy's operations and employees.

management team The individuals involved in managing a dairy operation.

manual scraping Removing manure from alleys using worker-operated equipment, such as a tractor or skid steer.

manure collection channel A channel running the width of an alley that collects manure as it is removed from the alley.

manure flow The physical pathways manure takes as it flows through a dairy site.

manure gas A lethal combination of hydrogen sulfide, carbon dioxide, and methane, given off by liquid manure in storage systems.

manure gutter A channel where manure is collected.

manure handling systems The systems designed to collect, store, and dispose of manure.

manure historical information Manure handling information regarding time frames, animal units involved, and amounts hauled.

manure separators Equipment used to separate liquid and solid manure components.

manure splashguard A guard designed to protect workers from being hit with manure while cows are being milked.

manure spreader Equipment used to spread solid manure on fields.

manure tanker Equipment used to spread liquid manure on fields.

marketing information Information provided to others regarding current and future projected values of milk, calves, cull cows, replacements, feeds, and so forth.

mastitis An inflammation or infection of the udder that causes soreness, hardening of the udder, and reduced milk quality.

maternity Housing for cows during the birthing process.

mattress-based stall A stall with a mattress-type surface, in which the mattress is composed of a fabric cover over an interior filled with soft material.

methane generation Using methane gases given off by liquid manure as a power source.

milk conductivity The electrical conductivity of milk, which is affected by sodium and chloride levels, used to identify cows with mastitic milk.

milk flow rate The level of milk produced at any given point in the milking of a cow.

milk let-down The ejection reflex caused by tiny muscle cells around a cow's milk glands, to squeeze milk out of the glands and into the milk ducts of her udder.

milk pasteurization Heating milk in order to kill unwanted organisms and bacteria.

milk per hour (mph) The total quantity of milk harvested in one hour.

milk per labor hour (mplh) The average quantity of milk harvested in one hour, divided by the average number of people milking.

milker An employee who milks cows.

milk-flow meter A monitor used to measure the rate at which milk is removed from the udder by a milking machine.

milking center All functional elements of a milking parlor, including the parlor, holding area, equipment room, and offices.

milking cow A lactating cow.

milking herd enterprise The portion of a dairy business that deals exclusively with the milking herd.

milking parlor A facility that cows are taken to and are milked in.

milking procedures A definition of all the steps that are required to take cows through the milking process.

milking routines The procedure that is followed to milk a group of cows on a specific dairy.

milking shift The time required to milk all of the animals in a herd once.

milking systems The system of equipment used to milk a herd of cows.

minimal prep A procedure used to prepare teats for milking that involves the minimum number of the udder-preparation steps recommended for proper hygiene.

MIRG (management intensive rotational grazing) An operation in which at least half of a herd's forage is harvested by grazing.

mission statement The statement of fundamental purpose and guiding philosophies of a business.

modern dairy management system A dairy system in which animals are housed in a freestall barn or open lots and taken to a separate milking parlor to be milked.

modernization The process of updating the technology used in a dairy's production, housing, and monitoring systems.

modified daily haul A system in which a dairy has sufficient storage to accommodate the manure produced in a few days, and relies on manure removal every few days rather than every day.

modified H-style A barn configuration in which the parlor complex is parallel to both freestall barns but located beside one of them rather than between them.

monitoring system The methods used to document and evaluate the production and financial performance of a business.

mound system A system used to house animals in open lots with dirt mounds. These mounds minimize mud because of the drainage provided by the slope of the mound and offer protection from wind during cold weather and access to breezes in warm weather.

natural ventilation The natural breezes that blow through a barn, optimized by correct barn orientation and elevation.

neck rail A metal rail designed to correctly position cattle in a freestall.

negative pressure The effect created by a fan that reduces the pressure within a barn, causing outside air to be pulled into an enclosed area.

net farm income The total amount of money remaining after all expenses are subtracted from all receipts for a certain period. A critical measurement of an business's profitability.

net merit An index used to evaluate the relative genetic merit of different animals, based on calculations that weight different traits according to their perceived importance.

net returns per cow The net farm income of a dairy divided by the average number of cows, which is an indicator of a business's profitability calculated on a per-cow basis.

NFIFO (net farm income from operations) A measurement of an operation's profitability that includes the profits resulting from the normal operation of the business for a specific period, but excludes any abnormal incomes that may be reported for that period.

nonpoint pollution source Indirect sources of pollution. Water contaminants that come from several different sources and are picked up and carried as water flows.

nuisance problem An annoying, unpleasant, or obnoxious situation; manure smells, for example.

nutrient budget A plan that is designed to determine the amount of nutrients that can be applied to land without allowing an unhealthy buildup of nutrients in the soil.

nutrient buildup The elevated level of particular chemicals in the soil as a result of excessive manure distribution.

ongoing support cost The cost that must be paid in order to maintain an activity.

open lot A dirt-based pen used to house animals.

open-ended questions Questions that encourage respondents to reply with anwers that demonstrate their knowledge and understanding.

open-sided barn A barn with no sidewall on one side.

operating procedures A formally defined series of steps to be followed to accomplish a specific operation on a dairy.

operator pit The area in which workers stand when operating a milking parlor.

organic product A differentiated product that meets the standards of the organic food industry.

organizational structure The administrative structure of a company, highlighting reporting relationships in particular.

outside feeding A method of feeding cattle at feed mangers located outside the barn.

owner's draw The amount of money an owner takes from the business for personal use.

pacesetting "Follow-me" leadership style, in which the leader likes to perform work-related as well as management activities.

palpation rail system A simple structure (normally constucted of pipes) that is used to restrain animals in a common area; also referred to as *management rail system.*

parallel parlor A style of parlor in which stalls are installed perpendicular to a parlor wall.

parlor capacity The total number of cows that can be milked in a parlor, given the desired milking frequency.

partnership arrangement An arrangement in which two or more people agree to co-own a business.

peak milk The point in a cow's lactation at which she produces her greatest quantity of milk per day.

pedometers Equipment used by EDDC systems to track the number of steps a cow takes in a given period of time.

people flow The physical pathways humans use to move through a dairy site.

people-passes Pathways through fences or gates that allow humans to enter or leave areas occupied by cattle.

percent equity A measure of solvency calculated by dividing the net worth of the person or business by the total assets owned.

performance standards and expectations Predetermined performance goals by which an employee can be evaluated.

pipeline milker A system that takes milk from individual milking units attached to cows where they are stabled and transports the harvested milk to a remote storage unit via a stainless steel pipeline.

plug-flow system A methane manure digestion system in which manure is added at one end of an enclosed container and allowed to slowly flow throught the container to an exit at the other end. As the manure flows through the container, the digestion process produces methane gases that are collected and used to generate electicity.

point pollution source Contamination of water that can be directly attributed to a specific cause.

post-calving The period of time immediately after a cow has given birth to its calf.

post-calving cow A cow that has just given birth to its calf.

pot-barn parlor A second milking parlor used to milk fresh and sick animals only.

pre-calving dry cow A cow that is expected to give birth to its calf in the near future.

predictive information Information based on observation and experience, used to plan future courses of action.

pre-dip milking hygiene Health practices performed in the parlor before milking units are attached (pre-dipping of teats with disinfectants, drying of the udder, etc.).

preliminary incubation (PI) count A measure of bacteria that will grow well at refrigerator temperatures.

prep-lag time The time between the beginning of teat preparation and the application of the milking machine.

prep time The time taken to manually clean, stimulate, and disinfect the teat surface.

prescriptive information Advice based on observation and experience, used to improve an operation.

prevailing wind The most common wind direction at a building site.

private contractor An individual hired to perform a specific task or service at a predefined cost.

production system The organized set of assets and procedures used to operate a business and produce a product.

profitability The returns remaining after all liabilities incurred to create a product have been met.

project management The act of monitoring a planned undertaking and ensuring that it is accomplished.

PTA fat Predicted transmitting ability of an animal for butterfat production, in pounds.

PTA milk Predicted transmitting ability of an animal for milk production, in pounds.

PTA protein Predicted transmitting ability of an animal for protein production, in pounds.

pulsator function The operational ability of a pulsator in a milking claw.

quarantine Separating infectious or newly acquired animals from others to reduce the potential spread of disease.

quick-release latch A locking device attached to gates, which allows them to be opened quickly.

rapid-exit stall Milking parlor stalls with front gates that are raised when a row of cows is done milking, allowing all cows in the group to exit at one time.

reception pit A pit that collects manure that has been scraped or washed down an alley.

recruitment The process of finding potential employees.

refusal That portion of feed delivered to animals that is not consumed.

relative feed value (RFV) An indicator of the quality of a forage based on its compositon.

replacement Another term for a *replacement animal.*

replacement animal An animal (usually a heifer) that is used to replace another animal when it is culled.

reporting relationships The hierarchy of authority in a company.

retirement planning Developing a plan to ensure that owners of a business are adequately provided for upon retirement.

retrofit A process of updating old facilities to be used for new functions or installing modern technology in old facilities.

return-on-assets A ratio that measures the rate of return on farm assets and is often used as an overall index of profitability. It indicates the profit returned for each dollar invested in the operation for the period of time being considered.

return-on-investments Another term used to refer to the return-on-assets.

RHA (rolling herd average) A measurement of the herd's average yearly performance over the previous 12 months. It is based on the average number of cows in the herd RHA number of cows, RHA milk, RHA fat, RHA protein, and RHA DIM (days in milk) are the most widely used values.

ridge opening Openings along the roofline of a barn that allow for air exchange in a naturally ventilated barn.

right of refusal The right to refuse a product due to substandard quality, guaranteed by a written provision in a contract.

robotic milking system See *AMS (automated milking system).*

roof pitch The pitch of a freestall barn's roof. It implies the number of inches of rise for each foot of barn width. The 4:12 roof pitch (four-inch of elevation rise for every twelve-inch of width) is recommended for dairy barns.

rotary parlor A style of parlor in which cows ride a rotating disk while being milked.

round-pipe flume system A large tube used to transfer manure by gravity from one location to another.

rubber mat A one-piece rubber-based pad installed in a freestall to provide cushion for an animal when lying, and traction when rising.

rump rail A rail installed behind cows in milking parlors to position them in the milk stall.

salmonella A pathogenic bacteria causing infection in cattle and humans.

sand separators Machines used to separate sand from sand-laden manure.

sand-based stall A freestall with sand used as the stall base and bedding material.

sand-laden manure Manure that contains sand, removed from sand-based freestalls.

scissor gate A gate that folds up out of the way when not in use, which functions as a normal gate when in the down position, and requires minimal space when in the up position.

scope of tasks and duties The range of actions and level of completion required for given tasks and duties.

S-corporation A special type of corporate business structure that allows profits to be taxed only once, as personal income. See also *C*-corporation.

sediment Fine debris that makes it way into milk as a result of dirty teats.

self-lock A head-lock that is activated by an animal when the animal enters the stanchion, resulting in the automatic restraining of the animal. Also referred to as "self-locking manger stall" or "self-locking stanchion."

sequential milking routine A routine in which operators split up the tasks of milking and work as a team, following each other and performing their individual tasks.

short-term goal Goals that are to be achieved in the near future.

shrink–swell potential The likelihood that soil will expand or contract as water is added to it.

silage bag A nonreusable plastic bag used to store forage.

silage leachate The liquid generated by the forage fermenting process, which flows from a silage container.

site elevation The height of a site in relation to adjacent land.

site selection The process of evaluating a parcel of land for its physical qualities and determining its acceptability for the placement of a dairy facility.

site survey The geological mapping of a parcel of land to determine its elevation profile, used to determine the amount of excavation required to prepare a site.

slatted (slotted) floor An alley floor with holes for manure to fall into a reception pit or storage tank below.

Social Security contributions Money paid to the government by both employees and employers, to fund the Social Security system.

social subsystem All elements of a dairy that are related to human interaction and satisfaction with the business.

sole proprietorship A business fully owned by one individual.

solids separator Machine used to separate manure into liquid and solid portions.

solvency The ability to meet financial obligations.

somatic cell count (SCC) A measure of somatic cells in milk, used as an indicator of mastitis.

sort gates Electronic (or mechanical) gates used to sort animals.

special-needs animal An animal with special health or reproductive needs.

special-needs barn A barn that houses animals with special reproductive or health needs.

springer A cow about to give birth.

springing heifer A heifer that has not yet calved but is close to having her first calf.

stack A method of storing silage on a flat surface.

staffing requirements The number and type of people required to complete all tasks on a dairy operation.

stall barn A dairy housing system that contains stalls where cows are tethered (tie-stalls, stanchions, etc.).

stanchion barn A method of housing in which cows are tethered in a stall using a stanchion.

standard operating procedure (SOP) A written document that explains exactly how a task is to be accomplished. Also referred to as procedure descriptions.

standard plate count (SPC) The total quantity of viable bacteria in a milliliter of milk.

steam-up The time period a few (normally three) weeks before a heifer or cow calves. The diet for steam-up animals is normally changed to prepare the animal for future milk production.

step-by-step approach The process of slowly growing an operation, in contrast to building a new operation all at once.

step-dam Gravity-flow manure channels used to collect and transport manure, constructed with several sections of flat floors that have a small dam at the end. The compartments formed fill with manure liquids, allowing the solid manure to float on the liquid, flow over the dam, and into the next section.

strategic decisions Long-term decisions.

strip Discharging a small amount of milk from an udder prior to attaching the milk machine, to check milk for mastitis and enhance the let-down process.

subway (basement) parlor Parlor in which the milk meters and some of the milking equipment are placed below the cows or the operators.

succession planning Planning for the transfer of an owner's interest in a business to other parties upon death or retirement.

sun angle The angle at which the sun's rays strike the earth, which varies seasonally.

super-hutch Housing stalls for post-weaning calves, typically holding between five and eight calves.

supplemental lighting Additional lighting provided to increase milk output.

supplier The provider of a particular product or service.

supporting The act of allowing employees to shoulder responsibility for a task but being available to offer assistance and advice.

swing parlor A style of parlor in which milking units are swung from one row of stalls to another for milking.

switch milking Use of an old stall barn to milk more cows than it is designed to hold. Groups of cows are milked in shifts. A 100-cow herd can milked in a 50-stall barn in two shifts.

tactical decisions Day-to-day decisions.

tail-to-tail barn A freestall barn designed with the rear of a row of cattle facing the rear of the next row.

teat condition The overall health of a cow's teats.

teat-end condition The health of the end of a cow's teats.

technical subsystem All elements of a dairy that are related to its facilities and equipment.

territorial milking routine A routine in which milkers are assigned units on both sides of the parlor and operate only the units assigned to them.

three-phase power A method of electrical power transmission in which three circuits are energized by alternating electromotive forces that differ in phase by one-third of a cycle.

tie-stall barn A method of housing in which cows are tethered in a stall using a strap around their neck.

TMR (total mixed rations) A blend of forages, grains, and supplements.

topographic survey The process of surveying the elevation and slope of a site.

topographical map The elevation map generated from a site survey.

total net return The return on an investment, after expenses have been subtracted.

total net worth A measure of solvency calculated by subtracting total liabilities from total assets.

total returns per cow A measure of financial returns per cow.

traditional dairy management system A dairy management system that relies upon older technology, which was common in most parts of the country before milking parlors and freestall barns were introduced.

traditional system See *traditional dairy management system.*

transition cow A cow in the last stages of pregnancy or the first stages of lactation.

transition difficulties Problems encountered bringing new cattle into an existing dairy.

treatment facility An area within a barn used to treat special-needs animals.

treatment-area-based animal handling system A method of handling animals by taking them to a designated treatment facility or special-needs barn.

trench silo A horizontal silo, lined with concrete or wood, in which silage is stored.

truck scale A scale mounted on a feed delivery truck, used to weigh feed.

T-style A barn configuration in which the parlor–holding area complex is perpendicular and attached to one freestall barn, with the second freestall barn behind and parallel to the first.

tunnel ventilation A ventilation system that uses fans placed at one end of a closed-sided barn to pull air through the barn and cool animals.

turn wheel The wheels at the end of a barn that hold the cables that pull an automatic alley scraper.

turns per hour (TPH) The average number of times a milking parlor is filled in one hour.

uddering up The udder of a dry cow or heifer enlarging before calving.

ultrafiltration Using filters to remove much of the water from milk to decrease the cost of shipping it long distances.

unemployment premiums Premiums paid to the government by both employees and employers to fund the unemployment insurance system.

unit attachment time The amount of time it takes to attach a milking unit to a cow.

unit on-time The amount of time that the milking machine is attached to a cow's udder as she is milked.

upright bin A vertical bin in which grain or concentrates are stored.

upright silo A cylindrical tower that holds feed.

vaccination program A biosecurity program put in place to ensure that all animals are vaccinated in an orderly fashion.

vacuum stability A measure of the variation in milk vacuum levels during the milking process.

volume price premium A financial incentive given to large dairies in exchange for receiving a large volume of milk.

walk-through flat-barn parlor A parlor installed in an old barn, in which cows exit through the stalls they were milked in.

waste disposal The method used to dispose of the manure generated by a dairy (normally applied to fields as fertilizer.).

water-flush A method of removing manure from alleys using a surge of water.

waterbed A freestall base surface composed of a rubber bladder filled with water.

wind shield An obstruction that blocks wind.

workman's compensation premiums Premiums paid by a company to a private insurance agent to cover employee injuries that occur during work.

World Wide Web (WWW) The international network of computer networks, which links individual computers to geographically dispersed resources.

Index

Note: Page numbers followed by "f" indicate figures and "t" indicate tables.

E

F

G

H

I

J

K

L

N

O